Water and the Search for Life on

David M. Harland

Water and the Search for Life on Mars

 Springer

Published in association with
Praxis Publishing
Chichester, UK

David M. Harland
Space Historian
Kelvinbridge
Glasgow
UK

SPRINGER–PRAXIS BOOKS IN SPACE EXPLORATION
SUBJECT *ADVISORY EDITOR* John Mason B.Sc., M.Sc., Ph.D.

ISBN 0-387-26020-X Springer Berlin Heidelberg New York

Springer is a part of Springer Science + Business Media(*springeronline.com*)

Library of Congress Control Number: 2005928955

Cover design: Jim Wilkie
Project Copy Editor: Alex Whyte
Typesetting: BookEns Ltd, Royston, Herts., UK

Other books by David M Harland

The Mir Space Station – a precursor to space colonisation
The Space Shuttle – roles, missions and accomplishments
Exploring the Moon – the Apollo Expeditions
Jupiter Odyssey – the story of NASA's Galileo mission
The Earth in context – a guide to the Solar System
Mission to Saturn – Cassini and the Huygens probe
The Big Bang – a view from the 21st century
The story of the Space Shuttle
How NASA learned to fly in space – the Gemini missions
The story of Space Station Mir

Creating the International Space Station
with John E Catchpole

Apollo EECOM – journey of a lifetime
with Sy Liebergot

NASA's Voyager missions – exploring the outer Solar System and beyond
with Ben Evans

Lunar Exploration – human pioneers and robotic surveyors
with Paolo Ulivi

Space Systems Failures
with Ralph D Lorenz

"The known is finite, the unknown is infinite; intellectually we stand on an islet in the midst of an illimitable ocean of inexplicability. Our business in every generation is to reclaim a little more land"
T.H. Huxley, late 19th century

"More public interest has in recent years been concentrated on the planet Mars than on any other"
Simon Newcomb, US Naval Observatory, 1903

"That Mars is inhabited by beings of some sort or other we may consider as certain"
Percival Lowell, 1906

"Surely one of the most marvelous feats of the 20th century would be the firm proof that life exists on another planet"
Stanley Miller and Harold Urey, 1959

"In all the history of mankind, there will be only one generation that will be the first to explore the Solar System, one generation for which, in childhood, the planets are distant and indistinct disks moving through the night sky, and for which, in old age, the planets are places, diverse new worlds in the course of exploration"
Carl Sagan, 1973

"Touchdown. We have touchdown!"
Albert R. Hibbs, Jet Propulsion Laboratory
as Viking 1 landed on Chryse Planitia on 20 July 1976

"An explorer would know, we have stood on the surface of Mars"
Tim Mutch, Viking Lander Imaging Team Leader

"Why explore Mars? The simple answer is, we're going to Mars to search for life. We're following the water because on Earth where you find liquid water, organic material and energy you find life"
Ed Weiler, NASA's Associate Administrator for Space Science, 2004

A successful search for extraterrestrial organisms would *"transform the origin of life from a miracle to a statistic"*
Philip Morrison, Massachusetts Institute of Technology

Patrick Moore
whose enthusiasm for astronomy has inspired so many

Table of contents

List of figures

Colour section, between pages 156 and 157

Author's preface

In 1965, with NASA's Mariner 4 bound for Mars, Patrick Moore published an updated version of his 1956 book *Guide To Mars*. Although a slim volume of just over 100 pages, it provided an account of essentially all that was known about the planet, complete with citations to the professional literature. It was the first book that I, aged ten, bought on astronomy. In 1992 the University of Arizona published a 1,500-page volume of papers written by in excess of 100 authors that summarised recent discoveries, and was soon out of date.

In discussing Martian geoscience, it is essential to adopt a strategy of 'multiple hypotheses'. In a science that has matured, it is possible to give many theories as if they are basic facts, as no one seriously questions them, but in a new field that has not reached this level of maturity almost everything is open to question, there are competing explanations, and, in the case of Mars, finding out which is correct is a wonderful process of exploration and discovery.

Establishing that life is not unique to Earth would, as the late Philip Morrison of the Massachusetts Institute of Technology pointed out, "transform the origin of life from a miracle to a statistic". Mars offers us the best chance of addressing this issue. There have been hints that life is extant today, but the evidence is disputed. However, if unambiguous evidence becomes available, there will be an almighty squabble to establish which previous data was – in retrospect – clearly evidence of that fact. If Mars is found *never* to have harboured life, the question becomes why did the chemical evolution that led to terrestrial life *not* do so on Mars. Irrespective of the outcome, the search for life on Mars addresses one of the most fundamental questions facing Mankind.

David M Harland
Kelvinbridge, Glasgow
June 2005

Acknowledgements

I would like to thank David Portree, Paul Spudis, Marc Rayman, Michael Hanlon, Alex Blackwell and Nick Hoffman for early comments; Pat Rawlings for the cover art; Corby Waste for other artwork; NASA, the Jet Propulsion Laboratory of the California Institute of Technology, Cornell University, Arizona State University, Malin Space Science Systems, the Royal Astronomical Society and Liverpool Astronomical Society for the illustrations; and of course, Clive Horwood of Praxis.

1

Peering at Mars

NAKED-EYE OBSERVERS

From his home on the island of Rhodes in the Aegean, Hipparchus, the greatest of the ancient Greek astronomers, compiled a catalogue of the positions and motions of the objects in the sky. He interpreted his observations as meaning that Earth was at the centre of everything, the planets circled around it, and the stars were fixed. Claudius Ptolemaeus, a fellow Greek living in Alexandria in Egypt a century later, observed that the planets did not precisely follow their predicted paths. However, since the circle was considered to be perfect, he proposed an 'epicycle' scheme in which, as each planet progressed around its orbit, it simultaneously traced a small circle around its mean position. Having studied mathematics at the University of Cracow, Nicolaus Copernicus realised in 1507 that the complexity of the epicycle scheme could be eschewed if it was assumed that the planets revolved around the Sun. Although Copernicus worked out his heliocentric theory in detail, it was not published until after his death in 1543, in the form of the book *De Revolutionibus Orbium Coelestium*.

In 1572 the Danish nobleman Tycho Brahe noted the appearance of a 'new star' in the constellation of Cassiopeia. It was as bright as the planet Venus for all of three weeks, then faded, and finally disappeared from sight a year later. When Brahe wrote up his observations in the book *De Nova Stella*, King Frederick II of Denmark was so impressed that he assigned a small island in the channel between Copenhagen and Helsingfors to Brahe to establish an 'observatory' from which to undertake a systematic study of the motions of the planets. As this was prior to the invention of optics, the observatory was provided with a live-in staff of assistants and a variety of instruments designed to accurately measure the positions of celestial objects. As his work progressed, Brahe hosted a series of visiting dignitaries and fellow scientists. However, when Christian IV assumed the throne he stopped the support, and in 1599 Brahe packed up his instruments and observations and relocated to Prague to continue work under the patronage of Emperor Rudolph II of Germany. On the death of Brahe in 1601, the unique archive of observations passed to Johann Kepler, who, in recent years, had been his principal assistant. At that time, the distinction between astronomy and astrology was so fuzzy that when Kepler set out to conduct a mathematical analysis of the motion of Mars, he subsidised his income by casting horoscopes.

As Kepler reported in 1609 in his *Astronomica Nova*, Mars's orbit is not, as had been thought, circular, it is elliptical, with an eccentricity of 0.093. As would become evident, Mars orbits the Sun about 50 per cent further out than does Earth – its heliocentric distance varies between 205 and 248 million kilometres, with a mean of 226 million kilometres. Earth, in contrast, never deviates more than 2.5 million kilometres from its mean heliocentric distance. As the planets pursue their orbits around the Sun in accordance with the laws inferred by Kepler, Mars, further out, travels through space more slowly. In terms of our calendar, Mars takes 687 days to make one revolution. Oppositions – when the planet is opposite the Sun in the sky – occur at intervals of 780 days. Although Mars will be at a different position along the Zodiac at successive oppositions, after eight oppositions – by which time Earth will have made 17 revolutions and Mars nine – it will have returned to the *same* part of its orbit. As a result, at perihelic oppositions, which occur only in the autumn, the two planets pass within 56 million kilometres of each other and Mars appears three times as bright as at an aphelic opposition in spring. In 1687 Isaac Newton published his *Philosophiae Naturalis Principia Mathematica* in which he expounded his law of universal gravitation, from which Kepler's laws of orbital motion could be inferred. And so, Brahe had provided a catalogue of exceptionally accurate observations without making any attempt to interpret them, Kepler had provided the empirical analysis without understanding why the planets move as they do, and Newton had identified the motivating force; this was a triumph for naked-eye astronomy.

TELESCOPIC STUDIES

Pioneers

In 1609, the same year as Kepler announced his discovery of the laws of planetary motion, Galileo Galilei in Padua turned a telescope to the sky and made a series of remarkable discoveries, such as: the Moon had a rugged surface; Jupiter had four satellites; and Venus showed lunar-like phases, *proving* that it circled the Sun. The next year, Galileo looked at Mars. His primitive telescope showed the disk but no detail. The first to report markings was Francesco Fontana, a lawyer in Naples, in 1636, but it is evident that the "black pill" that he placed at the centre of the disk was actually a flaw in the optics of his instrument, because he saw the same thing on Venus. More significantly, however, in 1638 Fontana drew Mars with a slightly gibbous phase (at its greatest departure from a circle, it is like the Moon about three days from 'full'). Neopolitan Jesuit Father Bartoli wrote of seeing two patches on Mars in 1644, but unfortunately did not draw them. G.B. Riccioli at the Collegio Romano, and his student F.M. Grimaldi, discerned albedo variations on the disk in 1651 and 1653, and particularly in 1655, when the planet was at perihelic opposition.

The first person to record what he saw on Mars's disk was Dutch lens-maker Christiaan Huygens. In late November 1659, after making a series of sketches over several nights and noting the times at which dark markings crossed the centre of the disk, he concluded that "the days and nights are about the same length as ours". This was the first discovery to be made of the nature of the planet, as opposed to its motion across the sky.

In Bologna in Italy, G.D. Cassini observed Mars when at opposition in March 1666, and made a series of drawings showing tremendous variation in the features. Although this was an aphelic opposition – meaning that Mars not well presented – he was nevertheless able to refine the rotational period. By observing over a series of nights, he noted that the planet rotated slightly slower than Earth. From the fact that it took 38 days for a given feature to appear at the same place on the disk at a given hour he was able to accurately calculate the rotational period as 24 hours 40 minutes. There are therefore 669 sols (as a Martian day has been termed) in the Martian year.

Huygens and Cassini both used refractors having small object glasses and long focal lengths for high magnification. Huygens relocated to Paris in 1666 to join the French Academy of Sciences at the invitation of Louis XIV. On accepting the first directorship of the newly founded Paris Observatory in 1669, Cassini installed a succession of ever more unwieldy refractors. Being in the same city, the two great planetary observers often collaborated. G.F. Miraldi, a nephew of Cassini, made a concerted study of Mars at its perihelic opposition in 1672, and concluded that many of the albedo features were clouds in a stormy atmosphere. In 1666 Cassini had seen a fuzzy white cap at the south pole but had not speculated as to its nature. Huygens had drawn it as a prominent feature in 1672. Miraldi noted a smaller cap at the north pole. In 1686 Huygens, having returned to Holland, made an 'arial' refractor with a lens 8.5 inches in diameter, but it was difficult to use because it had a focal length of 210 feet! Although he observed Mars at its 1688 perihelic opposition, he saw little more than he had using smaller instruments. Nevertheless, he continued to observe though to the aphelic opposition of 1694, the year before he died. At the perihelic opposition of 1704, Miraldi made a systematic study of the southern cap, and observed that it underwent a slight revolution, meaning that it was not centred exactly on the axial pole. He also noted that the areal extent of the cap varied over time. Cassini, who went blind in his later years, died in 1712. At the particularly close opposition of 1719, Miraldi, continuing his mentor's work in Paris, noticed that the southern cap disappeared for a time. This observation concluded the initial phase of the telescopic study of Mars, because, following Miraldi's retirement, the perihelic oppositions in 1734, 1751 and 1766 passed without comment.

The early refractors suffered chromatic aberration, and the wider the object glass the more severe this effect became. In 1663 James Gregory designed a reflecting telescope free of both spherical and chromatic aberrations, but his attempt to grind a parabolic mirror of glass was a failure. In 1668 Newton realised that a spherical mirror would eliminate chromatic aberration and, in view of Gregory's experience, made his mirror of polished speculum, an alloy of copper and tin. After making a model with a mirror 1 inch in diameter to demonstrate the idea, he made one twice as wide. Unfortunately, even when newly polished, speculum reflected only a small fraction of the light that impinged on it, and it rapidly tarnished. There is no record that Newton used his telescope for astronomical studies, and in 1672 he donated it to the Royal Society. However, in 1721 improvements in glass-grinding enabled John

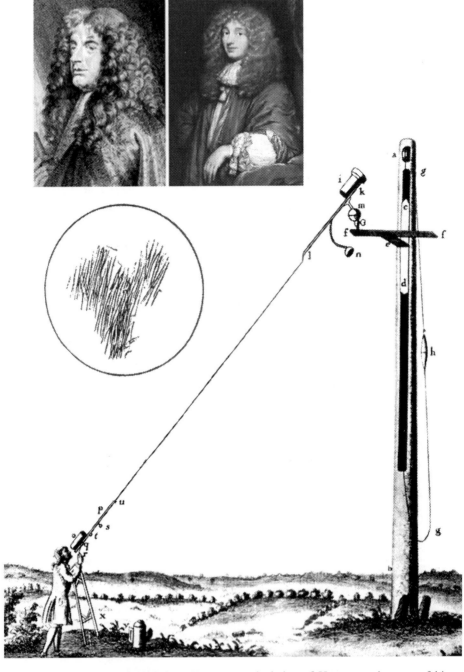

J.D. Cassini (left) and Christiaan Huygens, a depiction of Huygens using one of his 'arial' refractors and the drawing he made on 28 November 1659 of Mars's disk.

Hadley to create a parabolic mirror with a diameter of 6 inches, and produce a telescope capable of competing with contemporary refractors. In 1740 James Short began to manufacture such instruments for sale.

Herschel's observations

The accomplished musician F.W. Herschel took the post of organist at the Octagon Chapel in Bath in 1767, but his real passion was astronomy. By 1774 he was routinely giving eight music lessons during the day, then observing the sky at night. Having suffered the frustration of making and using long and unwieldy refractors, in 1774 he made a mirror 6 inches in diameter with a 7-foot focal length that proved superior to the telescopes used by his professional contemporaries. On 13 March 1781, while methodically seeking double stars in an effort to measure parallax, he noted a small sea-green disk that was not listed on his chart. Assuming this to be a comet lacking a tail, he monitored its motion against the stars for several nights and reported his findings. When the object was discovered to be a new planet, he suggested that it be named *Georgium Sidus* in honour of King George III, but continental astronomers objected and it later came to be known as Uranus. With an invitation to join the Royal Society and an annual allowance from the Crown, he was able to pursue his astronomical work on a full-time basis, and in due course a knighthood followed in recognition of his success.

William Herschel.

On occasions when Mars was well presented between 1777 and 1783, Herschel studied it using various reflecting telescopes, and discerned much more detail than his predecessors. At the perihelic opposition of 1781 he verified Miraldi's finding that the southern cap was not centred on the axis, and established that the northern cap was also misaligned. By monitoring the passage of dark features from night to night, he calculated the rotational period as 24 hours 39 minutes and 22 seconds. On two occasions in October 1783 he saw Mars pass in front of a star, and inferred from the rapidity with which the starlight diminished that the planet's atmosphere was thin. Nevertheless, in a paper to the Royal Society the following year he reported that, in addition to permanent surface detail, "I have noticed occasional changes of partial bright belts, and also once a darkish one in a pretty high latitude; and we can hardly ascribe such alterations to any other cause than the variable disposition of clouds and vapours." In fact, Herschel was able to integrate what was known about Mars, transforming it from a point of light in the sky into a world in its own right. Because its obliquity was similar to that of Earth, the planet had a seasonal cycle, and because its orbit was elliptical its velocity varied, being fastest at perihelion. Consequently, the four seasons were not of equal duration: the southern summer lasted 156 sols, but the southern winter lasted 177 sols. Since the planet received 44 per cent more energy from the Sun at perihelion than at aphelion, the fact that the south pole was tilted sunward at perihelion meant that the southern summer was hot, whereas the southern winter, with the pole tilted away at aphelion, was harsh. In the north, the temperature variation was much less dramatic and the duration of the seasons was reversed. By analogy with Earth, Herschel posited that the polar caps were fields of snow. As had Huygens, he presumed Mars to be inhabited, and ventured that "the inhabitants probably enjoy conditions analogous to ours in several respects".

In 1729, C.M. Hall in England found that positioning a concave lens behind a familiar convex lens would eliminate its chromatic aberration by bringing the light from the red and blue ends of the spectrum to the same focus. In 1674, by adding lead compounds to the casting, George Ravenscroft had created a 'flint' glass that had a high dispersion; therefore by making his 'corrector' from low-dispersion 'crown' glass Hall created a biconvex lens that was free of chromatic aberration. Previously, lenses had been given shallow curvatures to minimise the chromatic aberration, which had resulted in extremely long focal lengths, but a biconvex lens could be made with a stronger curvature for a dramatically shorter focal length – in fact, Hall's demonstration instrument, which had a lens 2 inches in diameter, had a focal length of 20 inches, in contrast to the more typical 20 feet. Although Hall did not make any effort to sell his lenses, John Dollond initiated production in 1750. Although the difficulty of casting flint glass disks for the main lens limited the aperture to approximately 4 inches, this led to a revival of refractors. In 1799, P.L. Guinand in Switzerland found that larger lenses could be made if they were stirred during casting to eliminate flaws, and in 1805 he teamed up with J. Fraunhofer to manufacture telescopes for sale. One of their 7.5-inch refractors made in 1812 outclassed Herschel's best reflector for planetary work. For the remainder of the nineteenth century, therefore, achromats were the instruments of choice.

Aerographers

While at the University of Berlin, J.H. Mädler, a child prodigy in mathematics, studied under J.E. Bode and J.F. Encke. In 1824 he teamed up with W.W. Beer, a wealthy banker with an interest in astronomy who founded a private observatory with a 3.75-inch Fraunhofer achromat, and in 1828 they set out to chart the Moon. They turned their attention to Mars at its perihelic opposition of 1830. As had their predecessors, Beer and Mädler pondered the character of the albedo variations. To assist in making accurate drawings, they used a micrometer to fix the locations of key points. By the time the opposition was over, they had a stack of fine drawings. They measured the rotational period to be 24 hours 37 minutes and 24 seconds. To resolve the discrepancy with Herschel's period, they re-examined his observations and found that over the interval of his study between 1777 and 1779 Herschel had overestimated the period by two minutes because the planet had rotated once more than he had thought. When this was corrected, their times agreed to within a few

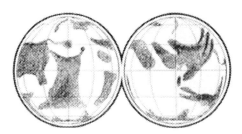

In 1840 W.W. Beer (left) and J.H. Mädler published the first map of Mars.

seconds. As for the albedo features, Beer and Mädler concluded that the planet's atmosphere was too thin to support a vigorous weather system. Confident that they had observed the surface, they drew a map, which they refined at each opposition. Soon after the map was published in 1840, Mädler left to take up the directorship of the Dorpat Observatory in Estonia, and Beer, who had not been the observer in the partnership, did no further work.

At the 1845 perihelic opposition, O.M. Mitchel of the Cincinnati Observatory, making one of the first studies of Mars from America, monitored the retreat of the southern polar cap and noticed that a narrow strip of its periphery was left behind and persisted for almost a month. This became known as the Mountains of Mitchel because it appeared to be a mountain chain whose summits remained freezing after the snow had melted from their lower slopes. In 1856 W. de la Rue in England, who had a 13-inch reflector, recorded more detail on the southern polar cap. The 1858 opposition was well studied by P.A. Secchi using the 9.5-inch refractor at the Collegio Romano's observatory. A short-lived white feature that he observed was the first sighting of what would later become known as a 'white cloud'.

Although perihelic, the 1860 opposition was too low in the sky to be studied from northern observatories but, in contrast, the opposition of 1862 was very well observed. Frederik Kaiser, director of the Leiden Observatory in Holland, produced a map that was a distinct improvement on that by Beer and Mädler. By comparing his timings with those of Huygens in 1672 and Herschel in 1783, he refined the rotational period to 24 hours 37 minutes and 22.62 seconds. Englishman William Lassell, a wealthy brewer, shipped his 48-inch reflector to the island of Malta to study Mars. In England, J.N. Lockyer used a 6.25-inch Cooke refractor to produce a

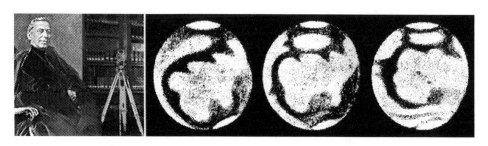

In 1858 P.A. Secchi made an extensive study of Mars.

J.N. Lockyer (left). William Lassell took his 48-inch reflector to Malta to observe Mars in 1862 (portrait courtesy of the Liverpool Astronomical Society).

set of exquisite drawings, which he later told the Royal Astronomical Society were in "marvellous agreement" with the map by Beer and Mädler. The dark features on Mars had generally been presumed to be seas, but in 1860 Emmanuel Liais – who had left the Paris Observatory to direct the Rio de Janeiro Observatory in Brazil – said that seas were unlikely to undergo a seasonal cycle of darkening and argued that they were *dry* sea beds in which vegetation bloomed in response to the retreating polar caps infusing water vapour into the atmosphere. Secchi disagreed, arguing that the seasonal waxing and waning of the caps "can be explained only by a melting of the snow or a disappearance of the clouds", and as liquid water was "a natural result of the behaviour of the snows", it was evident to him that "the existence of seas and continents . . . has been conclusively proved". John Phillips, who had observed Mars extensively during the 1862 opposition, said that if the dark areas were open water then he ought to have seen the Sun glint off it, but this had not been evident.

The Reverend W.R. Dawes, a friend of Lassell with an 8-inch Cooke refractor, made many drawings of Mars during the 1862 and 1864 oppositions. On Dawes's death in 1867, R.A. Proctor, a prolific author of popular books on astronomy, used these observations to compile a map that was considerably more detailed than that of Beer and Mädler. Proctor also *named* the features, using a nomenclature based on astronomers, living and dead, who had studied Mars. The prominent V-shaped feature first sketched by Huygens was named the Kaiser Sea. In keeping with the precedent set by Beer and Mädler, Proctor used the small dark spot on the equator – which they had labelled 'A', but he named Dawes Forked Bay – as the meridian against which to measure longitudes on the planet. In 1870 Proctor noted that he could recognise features on drawings made by Robert Hooke in 1666, and used this 200-year time base to refine the rotational period to 24 hours 37 minutes and 22.73 seconds. He also noted that, over that interval, an error of 0.1 second in the period would result in a discrepancy of two hours.

Camille Flammarion joined the Paris Observatory in 1858. A firm believer in life on other worlds, he published *The Inhabitants of Other Worlds* in 1862 and followed

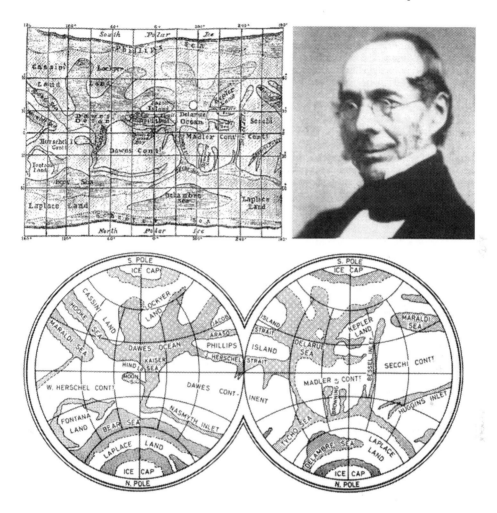

In 1867 R.A. Proctor issued a map of Mars (here shown in two formats) based on observations by W.R. Dawes (top-right) at the 1862 and 1864 oppositions.

up with *The Plurality of Inhabited Worlds* in 1868. He started observing Mars at its 1871 opposition. His map, published in 1876, retained the meridian, but with the reference feature more appropriately named the Meridian Bay. In 1878 he published *The Worlds in the Sky*. In 1883 he established his own observatory at Juvisy-sur-Orge, near Paris, and equipped it with a 9-inch refractor. In *Mars and its Conditions of Habitability*, published in 1892, he ventured that the planet had evolved more rapidly than Earth, its surface was heavily eroded and essentially flat, and the seasonal variations were inundations of the shorelines by seas "of Mediterranean shallowness". In *The Planet Mars*, in 1893, he recalled a speculation from the previous century by J.H. Lambert, and suggested the *ochre* areas were infested by plants of a species unique to Mars. "Why is not the Martian vegetation green?" he

In 1876 Camille Flammarion (left) published a map of Mars.

asked rhetorically. "Why should it be?" Arguing that "chlorophyll is made up of two compounds, one green, the other yellow", he ventured that "the yellow chlorophyll can exist alone, or be dominant" on Mars – a possibility that subsequently caught the imagination of storyteller H.G. Wells.

Aqueous vapour

In 1666 Isaac Newton was the first to realise that a glass prism refracts 'white' light into a rainbow of colours. In 1814 Fraunhofer noticed that the solar spectrum was crossed by dark lines, but their origin did not become evident until 1859 when G.R. Kirchhoff and R.W Bunsen realised that their positions corresponded to lines that appeared bright in the otherwise dark spectra of burning gases, and were thus in some sense characteristic of the substance present. In London, William Huggins likened this discovery to "coming upon a spring of water in a dry and dirty land". In 1867 he fitted a spectroscope to his 8-inch refractor and inspected the spectrum of Mars to determine the chemical composition of its atmosphere by the manner in which it reflected sunlight. On observing the planet when it was high in the sky, he saw the same aqueous vapour lines that he had seen in the solar spectrum when the Sun was near the horizon with its light passing through a long column of our own atmosphere. When the Moon had risen he inspected its spectrum and, on judging these lines to be absent, concluded that there must be aqueous vapour in Mars's atmosphere. Frenchman P.J.C. Janssen realised that the light bearing the signature of the planetary atmosphere would inevitably be modified on its passage through Earth's atmosphere, and that the observed spectrum would be a combination of the two. In 1867 he erected his spectroscopic telescope on the 10,000-foot summit of Mount Etna on the island of Sicily, Europe's tallest volcano, to minimise the local absorption. He made observations over two nights, first estimating the strength of the locally induced absorption by examining moonlight, and then waiting for Mars to rise in order to make his observation. He concluded that Mars's atmosphere was rich in water vapour. This result was verified by H.C. Vogel in Germany making similar observations in 1872, and by E.W. Maunder in England in 1875. However,

P.J.C. Janssen (top-left), H.C. Vogel, William Huggins and G.V. Schiaparelli. At his observatory in London, Huggins fitted his 8-inch refractor with a spectroscope.

visually inspecting the spectrum of moonlight, then waiting several hours for Mars to attain a similar elevation before making a mental comparison of the two spectra, left considerable scope for uncertainty, if only because local conditions might have changed during that interval. Nevertheless, the belief that there was water vapour in Mars's atmosphere reinforced the view that its polar caps were seasonal snow fields surrounding semi-permanent cores of water ice.

The canali

After graduating from Turin University in 1854, G.V. Schiaparelli studied under J.F. Encke in Germany and F.G.W. Struve in Russia. On his return to Italy in 1860 he joined the Brera Observatory in Milan, where he was appointed director in 1862. As Mars neared perihelic opposition in 1877 he used the observatory's new 8.6-inch Merz refractor to undertake a trigonometric survey in order to produce a refined map. He retained the accepted meridian but devised his own nomenclature, this time in Latin, and drew on terrestrial geography and classical literature for inspiration, as a consequence of which the Meridian Bay became Sinus Meridiani. Unlike his predecessors, Schiaparelli sketched the dark areas with sharp, rather than fuzzy boundaries. His initial map was primarily of southern latitudes, as that hemisphere was best presented, but at later oppositions he extended the coverage north and in 1888 issued a global map. The most prominent dark feature was Syrtis Major – which had been sketched by Huygens – off which led Sinus Sabaeus and, in turn, the other portions of the southern belt that subsequent observers would refer to as 'the great diaphragm'. In the northern hemisphere, the prominent Mare Acidalium was associated with a less-striking and less-regular belt. The area between was a vast ochre tract dotted with small dark patches such as Solis Lacus, Tithonius Lacus and Trivium Charontis. But the most astonishing feature of the map were the many narrow lines, which he named *canali*, meaning channels. "The canali run from one to another of the dark areas, usually called seas, and form a well-marked network over the bright part of the surface", he wrote, in presenting his map to the Royal Academy of the Lynxes in Rome. "Their arrangement appears to be constant and permanent. They traverse the planet for long distances in regular lines that do not at all resemble the winding courses of our streams. Some of the shorter ones do not attain 500 kilometres, but others extend for thousands of kilometres. Their number could not be estimated at less than sixty. Some of the lines are easy to see, others are extremely difficult, and resemble the finest thread of a spider's web drawn across the disk."

The 1877 opposition was also studied by E.W. Maunder using a 13-inch Merz refractor at the Greenwich Observatory, and while some of his drawings included features which, in retrospect, coincided with Schiaparelli's canali, these had not struck him as being worthy of note at the time. In fact Dawes had drawn several as ill-defined streaks, and there were also candidates on sketches by Beer and Mädler, Secchi, Kaiser and Lockyer. In an effort to refine Proctor's map, N.E. Green had set up his 13-inch reflector on the Atlantic island of Madeira, where the condition known as 'seeing' was excellent. Green criticised Schiaparelli for "turning soft and indefinite pieces of shading into clear, sharp lines". Making his own comparison of

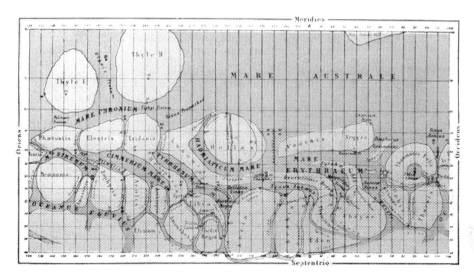

During the perihelic opposition of 1877 G.V. Schiaparelli drew a map of Mars's southern hemisphere with many of the dark features linked by narrow lines.

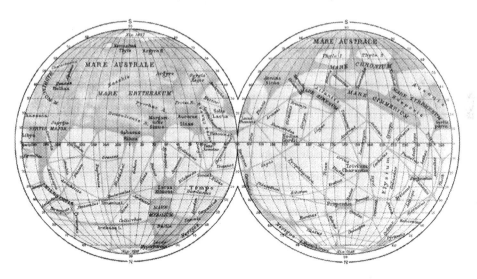

By 1888 G.V. Schiaparelli had extended his map to all latitudes and revised the projection.

the two observations, T.W. Webb ventured that, "much may be due to the different mode of viewing the same objects, to the different training of the observers, and to the different principles on which the delineation was undertaken." In particular, "Green, an accomplished master of form and colour, has given a portraiture, the resemblance of which as a whole commends itself to every eye familiar with the original. Schiaparelli, on the other hand, inconvenienced by colour-blindness, but of

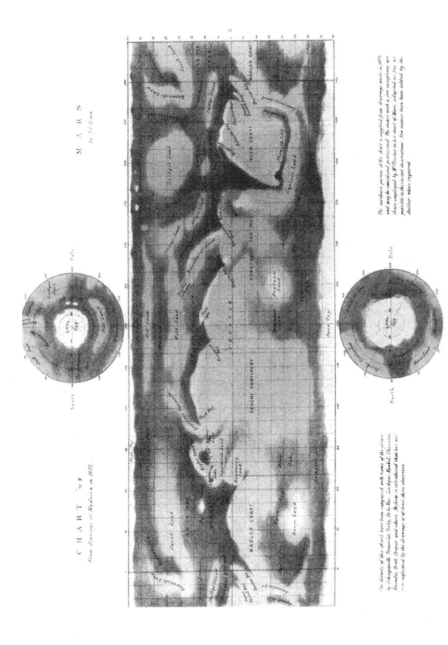

The map of Mars made by N.E. Green in 1877 in which such canali as he had observed were depicted as much softer features. (Royal Astronomical Society)

micrometric vision, commenced by measurement of 62 fundamental points, and carrying on his work with most commendable pertinacity, has plotted a sharply-outlined chart which, whatever may be its fidelity, no one would at first imagine to be intended as a representation of Mars." In other words, Webb decided, "one had produced a picture, the other a plan". But then, to be fair, Green was a professional artist while Schiaparelli was an engineer.

When Mars returned to opposition in 1879, Schiaparelli confirmed the overall character of the canali, and announced that a prominent one had 'geminated' (as he put it) into a pair of lines running in parallel, a short distance apart. He also noticed that some areas that he had drawn as featureless now showed a plethora of fine detail. In retrospect, it is evident that when Mars had been at perihelion two years earlier it had been partially obscured by dust storms. On observing a bright spot in the region he had named Tharsis, and presuming this to be a snow-capped plateau, Schiaparelli named it Nix Olympica – the Snows of Olympus. Green observed this opposition, this time from his home in London, although without significant result because the seeing was inferior to that of Madeira. The opposition of 1881–1882 was not so favourable because the planet did not come as close, but the seeing in northern Italy was excellent and Schiaparelli continued his study and recorded the appearance and disappearance, often over the space of only several days, of further geminations. Accepting that the style of his 1877 map was "purely schematic", he set out this time to make a map that was "more pleasing to the eye". For almost a decade, Schiaparelli remained the only person to claim to have seen the canali, but in April 1886 Henri Perrotin and his assistant Louis Thollon provided independent confirmation using the 15-inch refractor of the Nice Observatory, saying that they were "in nearly all respects, almost the character attributed to them". Even though that year's opposition was not very favourable, further positive reports came from A.S. Williams in England, H.C. Wilson in the United States, and F.J.C. Terby in Belgium, who inspected the planet "map in hand" to ensure that he knew where to look. However, many others, some using fine telescopes, continued to have no luck. Drawing canali soon became a popular activity, with people eagerly awaiting the next perihelic opposition in order to get a really good look at them.

In the meantime, there was a significant advance in telescope development. In 1856 the Italian astronomer C.P. Smyth had temporarily erected a telescope on the 12,200-foot summit of the volcanic island of Tenerife in the Atlantic, and reported excellent seeing. In 1874 James Lick, a Pennsylvanian piano and organ maker who had amassed a real-estate fortune in the Californian Gold Rush of 1849, awarded a grant of $700,000 for the construction of the world's largest refractor, and Alvan Clark was hired to manufacture a 36-inch-diameter lens. Lick died long before the telescope was commissioned in January 1888, but in keeping with his wishes his ashes were incorporated into the telescope's brick pillar. Situated on the 4,250 foot summit of Mount Hamilton near San José in California, the Lick Observatory was the world's first modern mountaintop observatory. The near-aphelic opposition of 1888 was unfavourable for Lick, but Schiaparelli was able to observe it from Milan using its new 19-inch Merz refractor, and said the canali still had "the distinctness of an engraving". However, by the near-perihelic opposition of 1892 his eyesight had failed. As Mars was still too far south in the sky for serious study by northern observers, W.H. Pickering of the Harvard College Observatory installed a 13-inch refractor at Arequipa, at an elevation of 8,000 feet in the Peruvian Andes. He drew the canali as hazy streaks, and in exceptional seeing was able to resolve small dark spots where canali intersected. In line with the prevailing belief that the large dark areas were seas, Pickering thought the small spots could be lakes. Since the canali had hitherto been seen only on the ochre areas, Pickering was astonished to discover a faint line crossing Mare Erythraeum. Realising that the dark areas could *not* be seas, Pickering adopted Liais's vegetative hypothesis, and suggested that the canali were swaths of vegetation living off volcanic gases that leaked from vast cracks in the crust of the otherwise inhospitable desert.

Martians

In 1638 a highly successful English businessman named Percival Lowell relocated to Boston in Massachusetts, and in 1855 the dynasty produced his namesake. The young Lowell graduated from Harvard University in 1876, and the following year was appointed to manage one of the family businesses. However, he was restless. Schooled in mathematics, Lowell had developed a passion for astronomy, and in 1893, having read Flammarion's books and corresponded with Pickering, he decided to establish his own observatory "to investigate the conditions for life on Mars" at the perihelic opposition of 1894. This would be no chimerical search, because, as he announced in advance, "there is strong reason to believe that we are on the eve of pretty definite discovery in the matter". English newspapers were so impressed by his aristocratic credentials that they erroneously referred to him as Sir Percival Lowell.

With Pickering's assistance, Lowell sited his observatory at Flagstaff, a small railroad town in Arizona which, by virtue of being on a plateau at an elevation of 7,200 feet, had excellent seeing for a large fraction of the year. Although time was short, an 18-inch refractor supplied by J.A. Brashear of Alleghany in Pennsylvania was ready for 'first light' on 23 April 1894. Starting on 28 May, Lowell, Pickering and their assistant A.E. Douglass studied Mars on virtually every night through the opposition to the following April, by which time they had compiled almost 1,000 drawings. As Lowell delightfully wrote, the canali were visible "hour after hour, day after day, month after month". As Schiaparelli had said that the appearance of the canali evolved through the observing season, and from one opposition to the next, Lowell was unperturbed that he could not find all of Schiaparelli's canali. On the other hand, he saw many more of them. In addition to mapping the small dark spots where canali intersected, Douglass verified Pickering's hitherto unconfirmed report of canali running across dark areas. Accepting that the dark areas were not seas, Lowell rejected the impression of Mars as "a second Earth" and instead saw it as an arid world on which almost all of the water was frozen into the polar caps.

On returning to Boston, Lowell mulled over his observations. He agreed with Pickering that the canali were swaths of vegetation, but rejected the idea that this formed alongside crustal fractures. Instead, he thought that the canali were laid out *purposefully*. He published his conclusions in early 1896 in a tome entitled simply *Mars*: "Firstly, that the broad physical conditions of the planet are not antagonistic

Percival Lowell established his own observatory to study Mars in 1894 and, like Schiaparelli, drew well-defined canali.

to some form of life; secondly, that there is an apparent dearth of water on the planet's surface and, therefore, if beings of sufficient intelligence inhabit it, they would have to resort to irrigation to support life; thirdly, that there turns out to be a network of markings covering the disk precisely counterparting what a system of irrigation would look like; fourthly, and lastly, there is a set of spots placed where we should expect to find the land thus artificially fertilised, and behaving as such constructed oases should." That is, the Martians lived where the canali intersected, at the dark spots on the ochre areas.

Lowell's view of Mars as a dying world was so evocative that it inspired H.G. Wells, in 1896, to write *The War of the Worlds*, a novel that began: "No one would have believed in the last years of the nineteenth century that this world was being watched keenly and closely by intelligence greater than man's and yet as mortal as his own; that as men busied themselves about their various concerns, they were scrutinised and studied, perhaps almost as narrowly as a man with a microscope might scrutinise the transient creatures that swarm and multiply in a drop of water. ... Across the gulf of space, minds that are to our minds as ours are to the beasts that perish, intellects vast and cool and unsympathetic, regarded this Earth with envious eyes, and slowly and surely drew their plans against us." By becoming an instant best-seller, this novel raised the public awareness of the Lowellian view of Mars. The astronomers, however, were not so impressed.

Early canali doubters

To many skilled observers, the canali were proving elusive. Having studied Mars in 1877 using an 8-inch refractor (only slightly inferior to that used by Schiaparelli) Henry Pratt reported to the Royal Astronomical Society that in exceptional seeing "glimpses were obtained of a structure so complicated and delicate that the pencil cannot reproduce it". In likening this to stippling, he continued, "what at first sight appeared as a broad hazy streak [could be] resolved into several separate masses of shading enclosing lighter portions full of very delicate markings".

At the opposition of 1892, C.A. Young pointed out that whenever the 9-inch refractor of Princeton's Halsted Observatory hinted at a canal, it was shown to be illusory using the 23-inch Clark refractor. That same year, E.E. Barnard discerned a small spot: "It is connected with the great sea south by a slender thread-like line. There is a small canal running north from Solis Lacus to a diffused dusky spot that does not appear on Schiaparelli's chart." But in 1896 he wrote that although the surface of Mars was "wonderfully full of detail . . . to save my soul I cannot believe in the canals as Schiaparelli and Lowell draw them. I see details where they have drawn none. I see details where some of their canals are, but these are not straight lines at all. When best seen, these canals are very irregular and broken up." When seeing was exceptional he saw a vast number of small faint markings that were so intricate as to be impracticable to draw. Some of the irregular dark streaks he saw matched the tracks of some of Lowell's well-defined canali. Significantly, he also saw irregular fine detail on the dark areas, supporting the growing belief that these were not shallow seas. Despite positive sightings of canali by people using small telescopes, Barnard opined that "before many oppositions are past" they would be seen to be "a fallacy". By now, having studied Mars using a variety of telescopes, Pickering had decided that the linear separation of the twin canals "was inversely proportional to the diameter of the telescope used, and directly proportional to the distance of the planet. In other words, if we use a telescope of twice the diameter we shall find the same canals will measure only half as many miles apart." Hence, "while the canals are undoubtedly genuine, their doubling is an optical illusion".

After studying in Berlin and Bonn, Vincenzo Cerulli returned to Italy, built his own observatory near Teramo, and equipped it with a 15.5-inch Cooke refractor in time to study Mars in 1897. When a prominent canal "lost its form of a line and altered itself into a complex indecipherable system of tiny patches" he realised that they were illusory. After two years of observing Mars on the limit of resolution, he was fascinated to discover that observing the Moon using low-power opera glasses revealed canali-like streaks that he knew were not distinct structures but fine detail that his sense of perception was linking. In 1903 E.W. Maunder came to a similar conclusion. Accepting that Schiaparelli drew what he had *seen*, Maunder argued that what Schiaparelli saw as a thin straight line was an interpretation of a mass of fine detail that was just beyond the resolving power of his telescope in that seeing. Maunder arranged with J.E. Evans, the headmaster of the Royal Hospital School in

E.E. Barnard at the eyepiece end of the 36-inch refractor at the Lick Observatory, here seen as it was circa 1900.

Greenwich, for a number of schoolboys to perform an experiment. The boys, who had no knowledge of the purpose of the test, were positioned at different distances from a disk bearing an impression of the light and dark areas of Mars, augmented by a plethora of fine dots. The nearest boys saw the dots, and drew them distinctly. Those further out saw only the general albedo features. Significantly, those on the limit of resolution for the dots drew fine lines. Lowell, of course, rejected what he referred to as "the small boy hypothesis".

The dismissal of Lowell's Mars
In 1894 Lowell had observed a broad blue band form as the southern polar cap retreated. First reported by Beer and Mädler, this phenomenon had been dismissed in 1892 by J.M. Schaeberle at Lick as an illusion induced by the contrast between the brilliant white cap and the adjacent ochre tract, but Lowell was sure that it was real. When Cowper Ranyard and G.J Stoney in Britain suggested in 1898 that the caps were a frost of frozen carbon dioxide, Lowell argued: "At pressures anything like one atmosphere, carbon dioxide passes at once from the solid to the gaseous state. Water, on the other hand, lingers in the intermediate stage of a liquid." But pressure was the issue. Unable to measure the pressure at the surface directly, he reasoned backwards: if the band was aqueous and the melt-water was regenerating vegetation, then this *required* the atmosphere to be sufficiently dense to maintain a moderate temperature. Despite the vast extent of the south polar cap in winter, the rate at which it shrank implied the ice was so thin that there was no more water on the planet than was present in the Great Lakes of America.

Lowell superseded the 18-inch telescope with a 24-inch Clark refractor in 1896 and continued to study Mars at successive oppositions, but with the range opening it was difficult to improve on his initial observations. Nevertheless, he refined his ideas, and in 1906 followed up his first book with the more provocatively entitled *Mars and its Canals*, in which he painted a more complete picture of the planet's inhabitants. The planet was being overwhelmed by a rapid process of desertification, and the polar caps were the only remaining sources of water. In a valiant effort to sustain their civilisation, the inhabitants had excavated the canals to direct the polar melt-water to the equatorial zones. His evidence was unequivocal. "That Mars is inhabited by beings of some sort we may consider as certain as it is uncertain what these beings may be." Although he offered no opinion as to their physical form, he developed a clear sense of empathy. The "cosmopolitan" scope of the canal network indicated a unified society. "Girdling the globe and stretching from pole to pole, the Martian canal system not only embraces their whole world, but is an organised entity. Each canal joins another, which in turn connects with a third, and so on over the entire surface of the planet. This continuity of construction posits a community of interest ... and the thing that is forced on us in conclusion is the necessary intelligent and non-bellicose character of the community which could thus act as a unit throughout its globe." Specifically, he argued that the Martians, faced with the extinction of their species, had banished warfare. "War is a survival among us from savage times and affects now chiefly the unthinking elements of the nation. The wisest realise that there are better ways for practising heroism, and other, more

W.H. Pickering (left), V.M. Slipher, W.W. Campbell and A.R. Wallace.

certain ends of ensuring survival of the fittest." (In contrast, less than a decade later humans launched their first 'world war'.) It was no accident that the canali were interpreted as artificial waterways. On Earth, canals were the state of the art in transportation systems. The Suez Canal had been completed in 1869, and the Panama Canal was under construction. It seemed reasonable, therefore, that the more advanced and more motivated inhabitants of Mars should have been able to build a canal network on a global scale. In *The Riddle of Mars*, published in 1914, C.E. Housden went so far as to explore in detail the engineering problems that the Martians must have overcome.

A.R. Wallace, the English naturalist whose work reinforced Charles Darwin's *On the Origin of Species by Means of Natural Selection*, and who published his own *Contributions to the Theory of Natural Selection* in 1870, was invited to review *Mars and its Canals*. He was so appalled that in 1907 he published *Is Mars Habitable?* in rebuttal. In particular, he took issue with Lowell's view that the temperature compared favourably with that of a summer's day in the south of England. Wallace argued that Mars must be much colder than this – in fact, it must be sub-zero – which meant that water could not possibly flow in open channels. "All physicists," he wrote in summary, "are agreed that, owing to the distance of Mars from the Sun, it would have a mean temperature of about –35°F even if it had an atmosphere as dense as ours. But the very low temperatures on Earth at the equator at a height where the barometer stands at about three times as high as on Mars, proves that from the scantiness of atmosphere alone Mars cannot possibly have a temperature as high as the freezing point of water," which, he insisted, was "wholly incompatible with the existence of animal life". His conclusion was that Mars "is not only uninhabited by intelligent beings such as Mr Lowell postulates, but is absolutely *uninhabitable*".*

---

* In fact, in criticising Lowell for using unscientific methods, Wallace erred in his derivation of the temperature on Mars and the abundance of water, but his analysis was better than Lowell's.

E.M. Antoniadi was invited to the Meudon Observatory to observe Mars using the 33-inch refractor. On his first viewing on 20 September 1909, he was astonished by the image clarity and saw immediately that the canali that Schiaparelli had depicted as continuous lines (bottom-left) were revealed as a multitude of spots (bottom-right).

To check on the reports by Huggins, Janssen, Vogel and Maunder of aqueous vapour, in 1894 W.W. Campbell at Lick had used a new spectroscope to compare Mars with the Moon, and found its spectrum to be "identical with that of the Moon in every respect". But in 1895 Vogel took photographic spectra, and announced: "It is definitely settled that Mars has an atmosphere whose composition does not differ essentially from our own, especially in being rich in aqueous vapour."

In 1902 Lowell had realised that when Mars was at quadrature, and its radial velocity was at its maximum of about 20 kilometres per second either towards or

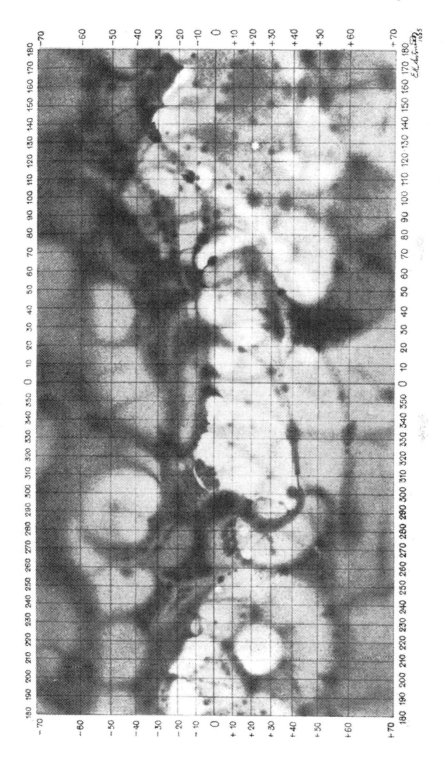

In 1925 E.M. Antoniadi published a map which emphasised the 'patchy' form of its albedo features, rather than their sharpness.

away from Earth, it should be possible to isolate its absorption lines from those imposed locally. In 1908, when the air over Flagstaff was exceptionally dry, V.M. Slipher took spectrograms of the Moon and Mars and concluded that the planet's absorption lines were slightly stronger, implying the presence of water vapour in Mars's atmosphere. Campbell, however, pointed out that the near-infrared line that Slipher had measured was on the steep decline in sensitivity of the film's spectral response, and thus open to doubt. As a prelude to the perihelic opposition of 1909, Lowell expanded on his theme in *Mars as the Abode of Life*, arguing that Mars's "smaller bulk has caused it to age quicker than our Earth, and in consequence it has long since passed through the stage of its planetary career which the Earth at present is experiencing, and has advanced to a further one, to which in time the Earth itself must come if it be not overwhelmed beforehand by other catastrophe. In detail, of course, no two planets of different initial mass repeat each other's evolutionary history, but in a general way they severally follow something of the same road."

In 1893 Flammarion hired E.M. Antoniadi as an observing assistant. That same year, Janssen founded the Meudon Observatory near Paris and equipped it with a 33-inch refractor with a lens made by P.P. & P.M. Henry. At the 1909 perihelic opposition Antoniadi was invited to use the Meudon refractor. His first inspection on 20 September was in excellent seeing, and he was astonished. "I thought I was dreaming," he wrote to Lowell. "The planet revealed a prodigious and bewildering amount of sharp or diffused natural irregular detail, all held steadily, and it was at once obvious that the geometrical network of canals Schiaparelli discovered was a great illusion." Although skilled at sketching, what Antoniadi saw "could not be drawn, hence only its coarser markings were recorded in the notebook". Campbell exploited this opposition to make another comparison of the spectra of the Moon and Mars, this time from the 14,500-foot summit of Mount Whitney, the highest peak in California. As Campbell was above much of the water vapour in Earth's atmosphere, and as the Moon and Mars were sufficiently close together in the sky to minimise other complications, he confirmed his conclusion that reports of water vapour in Mars's atmosphere were erroneous. However, because Campbell could take only a small instrument up the mountain he had been able to secure only low-dispersion spectra, and this enabled Lowell to dismiss his result. In 1910 Campbell gained higher dispersion spectra at Lick which proved conclusively that Mars's atmosphere is markedly deficient in both oxygen and water vapour. Although on Lowell's death in 1916 the public was still intrigued by his vision of Mars as being home to an ancient race of intelligent beings, the scientific community had dismissed him as a well-meaning crank. Nevertheless, despite Wallace's argument, the planet was still widely considered to host vegetation.

Surface speculations
Chlorophyll is green because it absorbs red and blue light, and reflects yellow and green. In 1918, and again in 1920, G.A. Tikhoff in Russia sought spectroscopic evidence of chlorophyll on the dark areas of Mars, but the expected absorption bands were absent. When it was realised that chlorophyll strongly reflects in the near-infrared, in 1924 Slipher, now director of the Lowell Observatory, took

photographs using an emulsion sensitive in this range, but the dark areas were even more prominent than at visual wavelengths. A.P. Kutyreva suggested that since Mars receives less than half as much sunlight as Earth does, Martian plants might have developed to absorb a wider range of wavelengths, and might *not* reflect the near-infrared. Y.L. Krynov examined a variety of plants and confirmed that some do use near-infrared energy, and argued that it was unreasonable to interpret the absence of strong near-infrared reflection as the absence of chlorophyll in the dark areas of Mars.

Although Wallace had dismissed Lowell's case for Mars being warm, both had based their arguments on indirect reasoning. It was not until the invention of the thermocouple in the early 1920s that it became possible to directly measure Mars's surface temperature. The first measurements were by E. Pettit and S.B. Nicholson in 1924 using the 100-inch Hooker reflector – then the largest such telescope in the world – at the Mount Wilson Observatory. They found that the midsummer noon temperature in the equatorial zone was around 20°C. In 1926 W.W. Coblentz and C.O. Lampland at the Lowell Observatory found the dark areas to be 10°C warmer than the ochre. Although a terrestrial observer cannot see the dark hemisphere of Mars because it is turned away, when the planet presented a gibbous phase it was possible to follow a given site across the terminator into darkness and measure the rate at which the temperature fell. Extrapolating from such measurements implied that the surface must plunge to –75°C (possibly –100°C) at local midnight. Such rapid radiation of heat to space implied that the Martian atmosphere was *very* thin. If the dark areas were due to vegetation, this would clearly have to be an extremely hardy strain.

Bernard Lyot at Meudon studied the polarisation of sunlight reflected by Mars at the oppositions during the 1920s and, after testing a variety of terrestrial rocks, he tentatively suggested in 1929 that Mars was volcanically active and blanketed by ash. W.S. Adams and T. Dunham used the 100-inch Hooker reflector in 1933 to repeat Lowell's suggestion of observing Mars at quadrature to distinguish between its spectral lines and those imprinted as the light passed through our atmosphere, and concluded that "the amount of oxygen in the atmosphere of Mars is probably less than one-tenth of one per cent of that in Earth's atmosphere over equal areas of surface". This did not mean that there was *no* oxygen on the planet, merely that if it was present it was at an undetectable concentration. Rupert Wildt at Princeton suggested that it had been drawn from the atmosphere and chemically bound in the surface material – the ochre was "strongly oxidised sandy formations, whose iron is almost completely in the form of ferric oxide". It seemed that Mars had rusted! In 1935 H.N. Russell noted that if Wildt was right, the planet must once have had a considerable amount of oxygen in its atmosphere, and since oxygen is reactive, it could have been maintained only if plants were present to replenish the supply – as occurs on Earth. In other words, while Mars may well be sterile today, as Wallace had argued, the presence of rust indicated that it must at some time have hosted life. This matched the popular belief that because Mars is smaller than Earth its interior must have cooled more rapidly and, being an 'evolved' world, it had been rendered sterile. If Lowell's race of Martians had indeed fought for survival as their planet dried up, they must have done so a *very* long time ago.

Surface pressure

While observing the perihelic opposition of 1939, Gerard de Vaucouleurs at the Le Houga Observatory in France made a determined study of how the brightness of a variety of albedo features varied as they moved across the disk in order to estimate the 'air mass' relative to viewing angle. In principle, because a site on the meridian was seen through a shorter 'column' than one near the limb, the way in which the surface appeared to vary in brightness as the planet turned provided a profile of the atmospheric density, from which he was able to calculate a surface pressure of 93 millibars. However, this technique was difficult to apply, and in the early 1940s other observers calculated values ranging between 112 and 120 millibars. In 1951 A.C. Dollfus of the Paris Observatory, using a polarimetry technique, estimated the pressure at the surface as 83 millibars. Overall, these results suggested that the surface pressure was about 10 per cent of that at Earth's surface. Using the 82-inch reflector of the McDonald Observatory on Mount Locke in Texas, in 1947 G.P. Kuiper made the first positive detection of carbon dioxide in Mars's atmosphere. Notwithstanding Campbell's negative detection of water vapour, in 1949 Kuiper reported that "the polar caps are not composed of carbon dioxide, [they] are almost certainly composed of water frost at low temperature, much below 0°C". Using the 24-inch refractor at Pic du Midi on a 10,000-foot peak in the French Pyrenees, Dollfus made similar observations in 1950, and drew the same conclusion. In 1954 de Vaucouleurs argued that although there must be residual cores of thick ice, the surrounding seasonal deposits were a frost of frozen dew in a coating that was only a few millimetres in thickness.

Bernard Lyot (left), Gerard de Vaucouleurs, G.P. Kuiper and H.C. Urey.

Lichens?

Having analysed the thermal data accumulated between 1926 and 1943, F. Gifford said in 1952 that, apart from the fact that it was considerably colder, the diurnal temperature at the Martian equator followed a curve that was very similar to that measured in the Gobi desert on Earth, implying that the physical characteristics of the regions were similar. Arguing against Wildt's 'rust theory', Kuiper pointed out in 1952 that the visible and near-infrared reflection spectrum of the ochre areas did not resemble that of the red rock and soil of the desert of the American Southwest. After comprehensive laboratory tests, he suggested that the closest spectral match was the brownish fine-grained mineral felsite (a potassic feldspar including grains of quartz) which implied that the crust had been subject to igneous processing. A polarimetry study led Dollfus to argue in 1956 that pulverised limonite (a hydrated iron oxide which, on Earth, forms mostly in or near deposits of oxidised iron) was a better match than felsite for the ochre areas. As for the dark features, E.J. Opik had argued in 1952 that these *must* be vegetation, otherwise the entire surface would have been rendered ochre by dust storms, and only the regeneration of plants which pushed up through the dust could re-establish the dark tone during the spring. In 1954, biologist H. Strughold said that the conditions on the surface of Mars were so inhospitable that only lichen seemed likely to survive. Strictly speaking, lichen is not a plant, it is a symbiotic association of a fungus and an alga, with the fungus providing an isolated environment for the alga and living off the wastes of the alga's photosynthesis. Lichens can live in environments in which neither the fungus nor the alga alone can survive, and have been found in the coldest of terrestrial niches. As algae and fungi are more primitive than 'higher' plants, it seemed plausible that they could have developed independently on different worlds. W.M. Sinton of the Harvard College Observatory used Lowell's refractor to secure infrared spectra of Mars in 1956, and proposed that faint absorption features near a wavelength of 3.5 micrometres were caused by acetaldehyde, and possibly even chlorophyll. Spectra by the 200-inch Hale reflector on Mount Palomar in 1958 and 1960 suggested these spectral features were present only in the dark areas. However, Sinton had used the Moon as a comparison, and the observations of Mars had been made on different nights from those of the Moon. In 1965 George Pimental of the University of California at Berkeley established that the spectral features were due to deuterated water – 'heavy water', in which one of the hydrogen atoms was replaced by a deuterium atom – in our own atmosphere.

Closest approach

After several decades of little progress in Mars research, the Lowell Observatory established a 'Mars Committee' in October 1953 to compile a reference book with the best available data on the planet and its atmosphere, as a preliminary to making a detailed study at the 1956 perihelic opposition. Additional interest was sparked by an event that was unrelated to the seasonal cycles. In 1954 an area the size of Texas rapidly turned dark, prompting D.B. McLaughlin of the University of Michigan to write the first of a series of papers arguing that Mars was intensely volcanic, and that the dark areas were blankets of ash. On studying the V-shaped streaks at the periphery of Syrtis Major, he said that they "point into the wind, and they are essentially point sources of some dark material that is carried from these points, fanning out because of variable wind direction". By analogy with Earth, the sources of these features "can have but a single interpretation: they are volcanoes". He proposed that the seasonal albedo cycles were due to the prevailing winds blowing ash across the ochre areas. The permanence of the dark features implied that the volcanoes were continuously active and that the wind patterns were reproducible year on year, otherwise, as Opik had noted, the windblown dust would soon mask

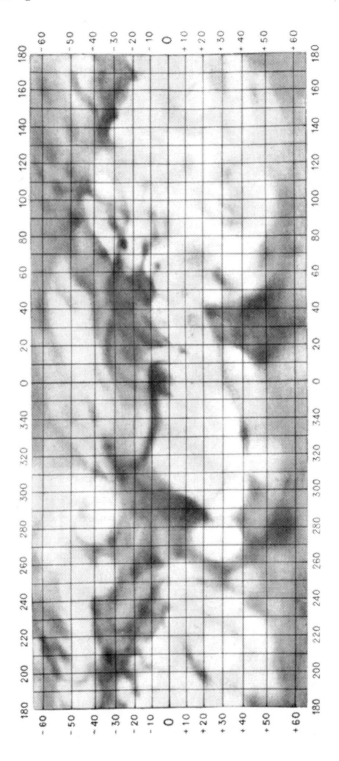

The albedo map (minus annotations) as defined by the International Astronomical Union in 1957. It represents the state of knowledge immediately prior to the dawn of the Space Age.

the dark areas. This theory attracted little support. When the planet was at its most gibbous phase, Lowell had sought evidence of mountains on the dark side of the terminator whose peaks were catching the Sun. However, this could not be done at opposition, when the planet was otherwise best positioned for viewing, because at that time the disk was 'full'. In 1952 Dollfus at Pic du Midi had tried and failed to detect mountains crossing the gibbous terminator. De Vaucouleurs agreed that there could not be any chains of mountains, and any isolated peaks could scarcely exceed a few thousand feet in height. It was hardly creditable, therefore, to imagine that the planet was rife with volcanoes. Another objection was that widespread volcanism would 'pump up' the atmosphere with gases, including water vapour, which was clearly not present. The theory also attracted criticism from those who believed Mars to have evolved more rapidly than Earth, as volcanism on Mars should long since have ceased.

Not only was the 1956 opposition perihelic, but Mars came as close as it ever does, and astronomers had hopes of using their new telescopes, instruments and techniques to make real progress, but in early September a dust storm started in the far south near the fringe of the polar cap, and within 10 days the entire planet was obscured and remained so for a month!

Atmospheric shock

The early optimistic reports of water vapour had been dashed by Campbell, and it was not until 1963 that vapour was finally positively detected. Dollfus set up a special spectroscope in the Swiss Alps in order to minimise the local absorption, to enable him to separate the planet's signal by its Doppler shift. Water vapour was only just measurable, and corresponded to a precipitable layer (i.e. the thickness of the layer of liquid water that would form if all the vapour condensed) of just 200 micrometres. At this same opposition, Hyron Spinrad at Mount Wilson used a new infrared-sensitive emulsion and calculated the precipitable layer to be a mere 14 micrometres – the planet was evidently drier than the most arid terrestrial desert. Nevertheless, liquid water would be able to exist on the surface at 85 millibars as long as the temperature did not exceed 35°C. For Project Stratoscope II in 1963, a 36-inch telescope with a spectroscope was sent aloft on a balloon to an altitude of 100,000 feet to observe Mars from above most of the water-laden troposphere, and it confirmed oxygen and water vapour to be present in trace quantities in the Martian atmosphere. As astronomer R.S. Richardson had ventured in his 1956 book *Man and the Planets*, "Two things [explorers]

In 1963 Project Stratoscope II sent a 36-inch telescope aloft on a balloon to an altitude of 100,000 feet.

would not need on Mars would be a fire extinguisher and an umbrella." In 1964 a spectroscopic analysis by Spinrad placed an *upper limit* to the surface pressure of 25 millibars, and put the *partial pressure* of carbon dioxide between 4 and 5 millibars. The remainder of the atmosphere was believed to be mainly nitrogen, but this was difficult to detect and in 1961 Harold Urey had made the radical suggestion that it was absent, which, if true, would make Mars so hostile than even lichen would not be able to survive.

2

A close look

SPACE TRAVEL

The rapid pace of rocket development following the Second World War persuaded many people to the view that space travel was imminent. In the early 1950s, in a series of articles in the New York magazine *Collier's* – magnificently illustrated by artist Chesley Bonestell – Wernher von Braun, who was arguably the world's most experienced rocketeer, explained how a fleet of massive nuclear-powered vehicles would ferry hundreds of people to Mars. Sceptics, however, considered that even the launch of small probes to investigate the planets was pure fantasy. In 1925 W. Hohmann, a German mathematician, studied the energy requirements for travelling from Earth to Mars, and found that the minimum energy 'transfer' to Mars involved departing from Earth on an elliptical orbit with an aphelion, achieved some 250 days later, that was tangential to Mars's orbit. The planet would have to be present when the spacecraft arrived, of course; so by calculating back from the time of interception and allowing for the planet's motion around the Sun, the time of departure could be identified. In fact, while the craft was in transit Mars would travel an angle of about 130 degrees. The best time to mount such a voyage would be when Mars was at opposition, and 'launch windows' were available that opened about 50 days prior to opposition and lasted for several weeks. This also, of course, depended on the rocket's capability. Because Mars has an elliptical orbit with the closest point of approach to Earth varying between 55 and 100 million kilometres, the ideal time to make such a flight would be at a perihelic opposition. Unfortunately, the necessary technology did not exist at the very close perihelic opposition of 1956, but having begun the Space Age on 4 October 1957 by launching Sputnik, the Soviet Union decided to dispatch two probes during the October 1960 window. Nikita Khrushchev had intended to announce the audacious undertaking in a speech to the United Nations in New York, but the rockets failed. Undaunted, another two space probes were prepared for the 1962 window. The first was stranded in Earth orbit by the failure of its 'escape' stage, but after being successfully sent on its way the second was named Mars 1; although its radio fell silent on 21 March 1963, the fact that it passed within 200,000 kilometres of its objective in June marked a great achievement. Meanwhile, NASA had attempted to send two spacecraft to Venus: a malfunctioning rocket destroyed Mariner 1 in July 1962, but Mariner 2 was successfully dispatched in August and during a 35,000-kilometre flyby on 14 December it became the first spacecraft to make *in-situ* observations of another planet.

Since Mars's opposition of March 1965 was aphelic, the Hohmann transfer was not favourable and the opportunity for supplementary telescopic studies was poor, but missions were still attempted. The Soviets had a new design of interplanetary craft. Launched in April 1964, Zond 1 was to make a flyby of Venus, but its radio soon fell silent. Zond 2 was dispatched towards Mars on 30 November and although contact was lost in April 1965 the trajectory was so accurate that it made a 1,500 kilometre flyby on 6 August that would have been perfect for imaging. Zond 3, which was launched on 18 July 1965, was not directed towards either planet: its trajectory took it within 10,000 kilometres of the Moon to test its imaging system, after which it simply passed into solar orbit to conduct a long-range test of its communications system.

FLYBY RECONNAISSANCE

NASA also exploited the 1964 launch window for Mars. Continuing the Mariner series, the spacecraft was constructed around an octagonal framework. Because the craft would be 3-axis stabilised in space, the elliptical dish of the S-Band high-gain antenna was affixed to the top of the frame in a position such that Earth would be in its narrow beam during the encounter with Mars. On the tip of a tube projecting along the axis of the dish was the low-gain antenna that would receive commands from Earth. Seven of the enclosed bays of the frame held spacecraft and scientific systems; the eighth held the liquid-propellant rocket motor that would make the mid-course correction. A cruciform of fold-down solar panels containing a total of 28,224 transducers would provide 300 watts of power at Mars's distance from the Sun. The attitude control system could tilt small vanes on the tips of the panels to correct any imbalanced forces imparted on the vehicle by the high-speed plasma of the solar wind. The deployed configuration spanned 6.8 metres and was 3.3 metres tall. Scientific instruments and associated systems accounted for 10 per cent of the vehicle's 260-kilogram mass. In addition to six particles and fields experiments to study the solar wind *en route* to Mars and in the planet's immediate vicinity, there was a system to image the planet during the flyby. Two identical spacecraft were built. In the case of Mariner 3, launched on 4 November, the aerodynamic shroud of the Agena-D stage failed to jettison. A timely modification enabled Mariner 4 to be launched on 28 November, a few days before the window closed. As soon as the craft was safely on its way, it hinged its solar panels down into the plane of the octagonal frame and oriented itself to face the panels towards the Sun to recharge its battery. Early in the interplanetary cruise, the star-tracker employed for attitude determination repeatedly lost its lock on the bright southern star Canopus and the solar plasma detector malfunctioned, but these problems were soon resolved.

There were several rules for Mariner 4's trajectory during the brief encounter. Firstly, neither Mars nor its two moonlets could be permitted to obstruct the craft's

An Atlas–Agena lifts off with Mariner 1.

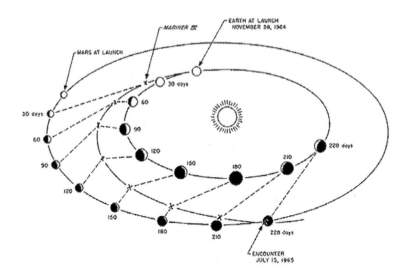

A schematic of Mariner 4's trajectory from Earth to Mars.

view of Canopus; nor could it pass through the shadows of any of these bodies. As the trajectory was designed to catch up with the planet in its orbit 240 days into the mission, the observations would be of the trailing hemisphere. The *post*-encounter trajectory was required to pass behind the planet (as viewed from Earth) in order to enable the manner in which the radio signal was cut off by the limb of the planet to yield the first definitive measurement of the density of the atmosphere. A second measurement would be made upon emerging from the leading limb. Crucially, this occultation had to be timed for when the planet was above the horizon of the radio receiver at the Goldstone tracking station in California. A mid-course correction on 5 December trimmed the initial 240,000-kilometre 'miss distance' down to the intended 10,000-kilometre flyby. In addition to arranging the required timing, this manoeuvre determined the surface features that would be in the camera's field of view at the time of closest approach. It had initially been hoped to get a look at the prominent dark feature Syrtis Major, but this was not now possible. The goal was to secure pictures comparable in resolution to contemporary telescopic pictures of the Moon.

On the base of the octagonal frame was a vidicon imaging system carried on a platform that could be rotated through 180 degrees. The system was powered up 6 hours prior to the encounter on 14 July 1965, to allow time to troubleshoot any problems – diagnosing a fault and issuing recovery actions required to be prompt, as the light-speed travel time over such a distance was 12 minutes. With 1 hour to go, the platform rotated until its wide-angle sensor noted the planet's presence, at which time it centred the camera on the planet. The imaging sequence was started when the narrow-angle sensor detected the planet, about half an hour later. (As a contingency measure, the imaging system, which was 'looking', would have self-triggered as soon as the planet's disk encroached on its field of view.) A television camera displayed the image from a 4-centimetre-diameter aperture f/8 Cassegrain telescope with a

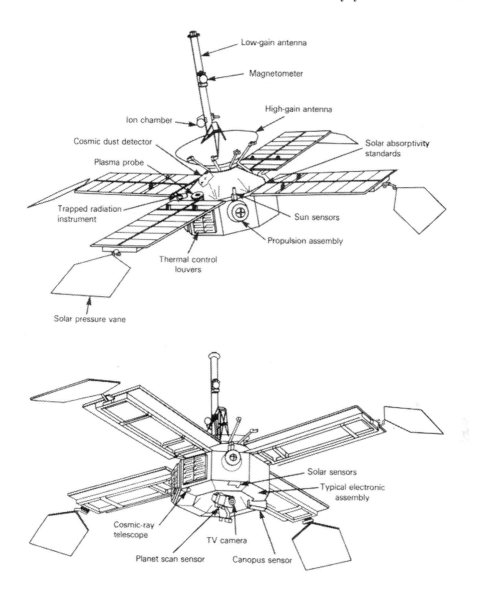

A schematic of the structure and instruments of Mariner 4.

focal length of 30 centimetres onto a small phosphor screen, and this was digitised into an array of 200 by 200 pixels, each of which encoded a 6-bit greyscale value. The data from the particles and fields instruments was transmitted in real time, but the images were stored on a 100-metre loop of magnetic tape. The exposure time had been pre-set at 1/20th of a second – this being the best estimate on the basis of predicted illumination. As it took 24 seconds to read out the image from the vidicon, and another 24 seconds to clear the screen, pictures could be taken no more rapidly than once every 48 seconds. A rotating shutter alternated blue–green and orange–red filters in order to enhance the contrast in the greyscale and to emphasise the colour differences of the albedo features on the Red Planet. About 12 hours after the encounter, the tape began to replay the images. At 8.33 bits per second with the 10-watt transmitter, it took 8 hours and 20 minutes to send the 240,000 bits in each image. In fact, the effective rate was one frame every 10 hours because each frame was accompanied by engineering data.

Taken from a range of 16,500 kilometres, the first frame showed the planet's limb and the blackness of space. The engineers were ecstatic, since this confirmed that the camera had properly acquired the planet. As the other frames were processed there was disappointment. Despite having used filters to highlight the contrast on the planet's surface, most of the 22 images were bland due to 'flare' in the optics, and the remainder were black. Fortunately, the Jet Propulsion Laboratory (JPL) – a part of the California Institute of Technology in Pasadena, near Los Angeles – which had developed the spacecraft, had a computer algorithm to 'enhance' digital imagery by adjusting the shades of grey to 'stretch' the contrast range, and when this was applied some surface details became evident. The imaging sequence began with a view looking northwards across the limb at a longitude of 190 degrees, and tracked southeast and across the equator at longitude 180, to about 52 degrees south, where it swung north and crossed the terminator having spanned some 90 degrees of longitude. In the first few frames the surface appeared to be blotchy, with hints of large circular features. On frame number 7 it became evident that these were craters, which were depicted in greater clarity with each successive frame. The crater 120 kilometres in diameter on frame 11, taken from a range of 12,500 kilometres, immediately became the iconic Mariner 4 image (and was subsequently named 'Mariner' in honour of the craft). The final frame on which surface detail could be seen was number 15. Although the contrast on the terminator would have emphasised the surface relief, the sensor that was to have increased the exposure as the field of view darkened failed, with the result that the final few frames were black. Although the imaging system was marred by flare, it *had* achieved its primary objective of revealing the nature of the Martian surface in a swath covering about 1 per cent of the planet. The fact that craters were present on a variety of albedo features implied that the entire surface was *billions of years* old. That the craters had retained their shapes implied that the planet was not subjected to the tectonism that continually reshapes Earth's surface. Furthermore, the state of preservation of the craters implied that Mars was never wet enough for a hydrological cycle in which the surface was chemically and physically eroded by rainfall.

The final 'nail in the coffin' for Martian life was delivered by the occultation experiment. The strength of the vehicle's radio signal was carefully monitored as it flew behind the planet's limb an hour and a half after the point of closest approach and again as it emerged from the far side an hour or so later, in order to produce a 'refractivity profile'. Because different mixtures of gases have different refractive effects on a signal, this data enabled the model of the atmosphere to be refined in terms of chemical composition, temperature and pressure as functions of altitude in daylight at 55 degrees southern latitude and in darkness at 60 degrees north. The

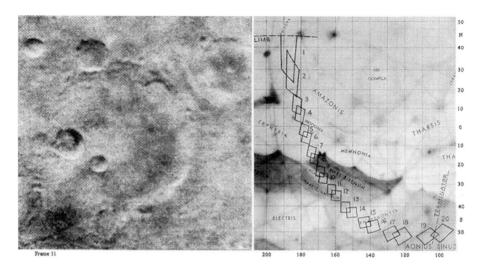

The outlines of the images taken by Mariner 4, superimposed on a section of an albedo map drawn by telescopic observers, and frame number 11 showing the 120-kilometre-diameter crater that was named in honour of the spacecraft.

surface pressures were 4–6 millibars. Considering the discovery by Mount Wilson that the partial pressure of carbon dioxide at the surface was 4 millibars, the occultation data indicated the atmosphere to be at least 95 per cent carbon dioxide. By analogy with Earth, it had been presumed that the remainder of the '25 millibar atmosphere' was predominantly nitrogen, but this gas was actually present only in trace amounts. Liquid water would certainly not be stable at such a low pressure. Furthermore, Mars was much colder than had been expected, because after sunset the surface would soon radiate the heat it had absorbed from the Sun. Although it had little water vapour in absolute terms, the fact that the planet was so cold meant that at night the atmosphere was close to its saturation point. The discovery that the poles chilled to temperatures below $-128°C$ revived the long-discarded idea of the seasonal caps being a frost of carbon dioxide rather than water ice. Finally, it was realised that the low pressure was not a coincidence: the carbon dioxide in the atmosphere and the caps was in a state of equilibrium. The other instruments established that if Mars had a magnetic field it was exceedingly weak, and unable to form a magnetosphere to ward off the plasma of the solar wind, and the fact that the surface was irradiated by solar ultraviolet and energetic particles from the solar wind rendered most unlikely the prospect of life even as 'simple' as lichen.

In view of the tenuous atmosphere and the seemingly inert ancient surface, *The New York Times* made a play on Mars's moniker of the Red Planet by dubbing it the Dead Planet, and, almost overnight, scientific interest plummeted.

Confirmation

NASA was too busy preparing for the Apollo missions to exploit the next window for Mars, but it decided to mount two flyby missions in 1969. The imaging system

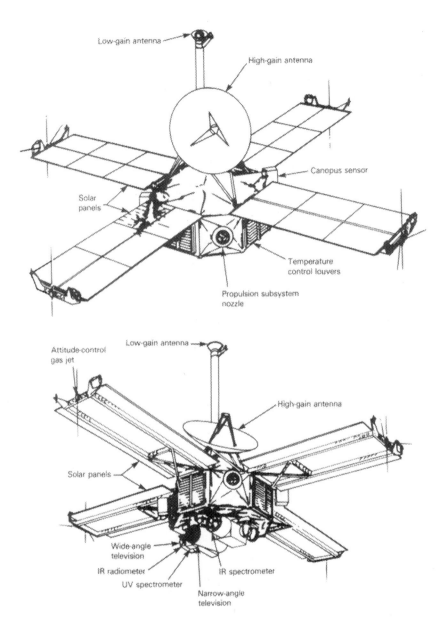

A schematic of the structure and instruments of Mariners 6/7.

was augmented with a wide-angle lens to provide context for the narrower frames and mounted on a more sophisticated platform, the tape recorder was upgraded to store more images, and the communications system was upgraded to transmit them at a rate 2,000 times faster, at 16 kilobits per second.

On 29 July 1969 Mariner 6 performed its far-encounter task by taking a series of

50 pictures in order to document the sunlit hemisphere of the planet as it turned on its axis. Even the first picture, taken from a range of about 1 million kilometres, had better resolution than that of a terrestrial telescope. With these transmitted and its tape recorder reset, the following day the spacecraft took 25 pictures along an equatorial swath running from 320 to 60 degrees of longitude, the point of closest approach being 3,500 kilometres above Sinus Meridiani. On inspecting the early results, the scientists decided that the far southern latitudes were so intriguing that they hastily reprogrammed Mariner 7 to take an additional eight pictures to extend its near-encounter coverage into that region. Mariner 7 began its far-encounter sequence three days later, taking 91 pictures. There was some concern when it fell silent as it neared the planet, but contact was re-established and it took 33 images during its flyby on 5 August. It documented two swaths, one crossing the equator and running from 100 to 20 degrees longitude, and the other in the southern hemisphere close to the meridian. The occultations confirmed that the pressure profile of the atmosphere is shallower since Martian gravity is weaker. The surface pressure at Sinus Meridiani was 6.5 millibars, but the fact that the southern latitudes were 3.5 millibars implied that this was elevated terrain. It was decided to utilise 6.2 millibars as a convenient datum. This pressure is significant because it is the 'triple point' of water. In fact, given geographical and seasonal variations, the *mean* surface pressure is near this value. Liquid water is unstable at lower pressures. The suggestion that the polar caps were frozen carbon dioxide was strengthened by Mariner 7's measurement of the temperature at the south pole as $-123°C$, because this was close to the carbon dioxide 'frost point'. Interestingly, the floors of some of the craters on the fringe of the cap were coated with frost.

Between them, these three flyby missions revealed about 10 per cent of Mars's surface at moderate resolution. It became evident that craters were not ubiquitous.

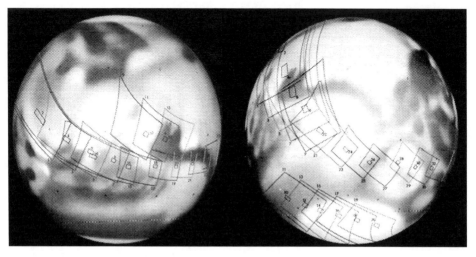

The outlines of the images taken by Mariners 6/7, superimposed on an albedo map drawn by telescopic observers.

Photogeologists identified 'cratered', 'chaotic' and 'featureless' terrains (in order of most-common to least-common). The morphology of the chaotic terrain south of the equator at 40 degrees longitude suggested a general collapse, implying endogenic activity, but what could have eroded the subsurface on such a scale? A bright circular feature called Hellas, which telescopic observers had seen brighten to resemble an offset polar cap, had been assumed to be a plateau that attracted a cover of snow in winter, but was revealed to be an enormous depression and – in striking contrast to the adjacent cratered terrain – appeared to be featureless. What could it be?

Thoughts on the nature of the surface
In the early 1960s Carl Sagan and James Pollack at Cornell University argued that if, as Dollfus had suggested, a lot of limonite was exposed at the surface then this might be being eroded to dust. The newly exposed coarser limonite would be relatively dark. If the seasonal winds were cyclic, this highly reflective dust would be blown off the high ground in the spring and collect in low-lying areas, then be blown back in the autumn. In this theory, the albedo variations were not the result of chemical variation of the crust, rather the erosion of a *ubiquitous* mineral. This was an appealing explanation of the seasonal variations since it did not involve the action of water vapour. In 1966 Sagan and Pollack analysed the data from back-scattering radar studies using radio telescopes. It was immediately evident that the reflection was strong when the beam illuminated a bright area, and weaker when it illuminated a dark area. This was interpreted as meaning that the albedo variations related to the physical character of the surface. The data did not directly measure altimetry, but Sagan and Pollack had already inferred from their limonite analysis that dark areas were elevated and bright areas were low-lying. It appeared, therefore, that the low-lying land was thick with dust and the elevated dark areas were rocky outcrops that were periodically partially masked by windblown dust. The elevation range was calculated to span 10 kilometres, but the features were so broad that the slopes were shallow. This contradicted the traditional view from telescopic studies that the dark areas were low-lying areas of vegetation which, although periodically partially overrun by dust, was able to reassert itself. Several years later, however, Gordon Pettengill of the Haystack Observatory of the Massachusetts Institute of Technology timed radar reflections to measure altimetry, and found that there was *no relationship* between albedo and elevation. The bright Elysium was high, but so was the dark Syrtis Major. The most elevated area was the bright Tharsis region. The early radar work was confined to the equatorial zone, but a later study showed that the bright circular Hellas was the lowest-lying region.

A GLOBAL PERSPECTIVE

The flyby missions had repudiated the soundly reasoned impression of the Red Planet, turning the expected cold dry world that harboured vegetation into one that still bore the scars of ancient impacts and had an atmosphere too thin for liquid water on the surface, water-ice, or life. Nevertheless, geoscientists were eager to place a vehicle into orbit to map the planet. In fact, it would be better to send two spacecraft. When we see the Moon at its 'full' phase, we see only albedo variations. At other phases, shadows along the terminator highlight the surface relief. No telescopic study of Mars had ever hinted at topography on its terminator, which was why attention had focused on albedo features. A spacecraft in orbit would be able to view the surface under a variety of illuminations, but whereas a high point over the illuminated hemisphere would favour albedo studies, a low point over the terminator was required for topographic mapping. NASA therefore decided that Mariner 8 would adopt a highly inclined orbit that dipped low over the terminator in order to map about 70 per cent of the planet at high resolution, while Mariner 9 adopted an almost equatorial orbit at high altitude to monitor the seasonal albedo variations. The improved cameras would be mounted on scan platforms that could nod up and down as well as rotate axially. Since these new spacecraft would have to incorporate a rocket and the propellant to brake into orbit, they would be much heavier than the flyby spacecraft, but the 1971 perihelic opposition would minimise the energy requirements. Unfortunately, the Centaur stage of the launch vehicle for Mariner 8 failed. The mission was hastily redesigned to enable a single spacecraft to make as many observations as possible from a compromise orbital inclination of 65 degrees that would provide illumination at shallower angles than would be ideal for albedo studies and higher than the ideal for topographic studies. To everyone's relief, the reprogrammed Mariner 9 was successfully dispatched on 30 May 1971.

Telescopic observers had noticed that 'yellow clouds' sometimes masked large areas of the Martian surface. In 1909 E.M. Antionadi had seen one that lasted for several days. In 1911 he had seen another that rapidly expanded to engulf much of the southern hemisphere for several months. When such obscurations were seen in 1924, 1929, 1941 and 1956, it became evident that they were dust storms triggered by strong winds caused by differential solar heating in the southern hemisphere at perihelion. In February 1971 C.F. Capen of the Lowell Observatory predicted that a major dust storm was likely to develop during that year's opposition and warned that this might interfere with the forthcoming orbital mapping mission. On 21 September a storm did indeed develop. It was first photographed by Gregory Roberts using the 27-inch refractor at the Republic Observatory near Johannesburg in South Africa, and by 27 September it had obscured a wide area west of Hellas. By the end of the first week of October the mid-southern latitude zone was masked and by the end of the month the entire planet was obscured! When Mariner 9 began its far-encounter imaging on 10 November, one week out from the planet, the only features were the south polar cap and four fuzzy dark spots in the Tharsis area, three of which were spaced about 700 kilometres apart on a line running southwest to northeast across the equator, with the fourth off to the northwest. Although the trio were promptly dubbed neutrally as 'North Spot', 'Middle Spot' and 'South Spot', their positions corresponded to Ascraeus Lacus, Pavonis Lacus and Nodus Gordii respectively, all of which were believed to be low-lying. The fourth matched Nix Olympica, which the Mariner 6/7 far-encounter imagery had suggested marked a large impact. How could depressed areas be visible through the dust storm? Astronomer Brad Smith, leading the imaging team, recalled that when he had

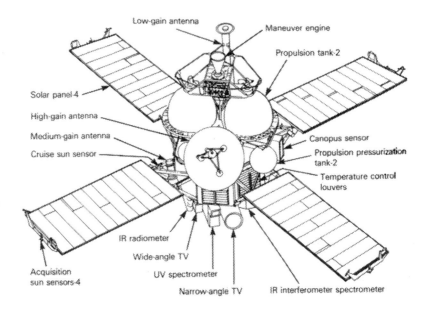

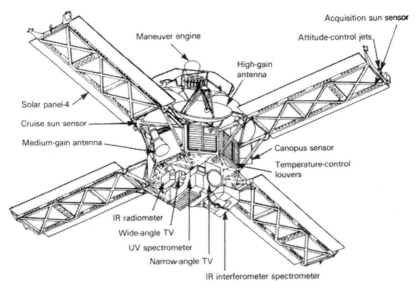

A schematic of the structure and instruments of Mariner 9.

photographed the planet in violet light while its atmosphere was clear, these four features had been brilliant when approaching the dusk terminator, suggesting clouds of water-ice around the summits of mountains. Was Mariner 9 seeing the summits of very high mountains protruding out of the dust storm?

On 14 November Mariner 9 fired its braking rocket and entered Martian orbit.

An Atlas–Centaur lifts off with Mariner 9.

Each time the craft crossed the limb, its radio signal was monitored to characterise the dust-laden atmosphere. Meanwhile, it was told to train its cameras on the dark spots. As the dust started to clear, a large circular structure emerged in each case.

Were they impact craters on a high plateau? When the spots turned out to have a nested structure, it was realised that they must be vast calderas on the summits of volcanoes. When the dust cleared, Olympus was seen to be a shield volcano rising 25 kilometres above its surroundings, and its base was marked by an 8-kilometre-tall scarp. The other spots were also found to be volcanoes with summit calderas. In contrast to the impression gained from the cratered terrain revealed by the flyby probes that Mars had never been active, it was clear that it had undergone intense volcanism. Although the spacecraft's infrared radiometer saw no 'hot spots' in the calderas to indicate activity, high-resolution views of lava flows on the flanks were sufficiently lacking in impacts to suggest that the vents had been active in 'recent' times. Although tenuous and dry today, the atmosphere might from time to time be enriched with gases, including water vapour, from volcanic eruptions.

When the dust settled in January 1972, Mariner 9 refined its orbit to initiate mapping, which began in the southern hemisphere and expanded northward, with a series of startling discoveries. There were sinuous channels in the cratered terrain, some with a dendritic form. As the coverage crossed the equator, a canyon system was revealed running one-quarter of the way around the planet. Further north, vast flood channels drained into the Chryse region, which was part of a low-lying plain that formed much of the northern hemisphere. It seemed that, after all, Mars must once have supported a hydrological cycle, which indicated that it had undergone a major change in climate. "It soon became apparent," noted Jack McCauley of the geology group, "that almost all generalisations derived from Mariners 4, 6 and 7 would have to be modified or abandoned. The participants in the flyby missions had been victim of an unfortunate happenstance in timing. Each earlier spacecraft – except in part for Mariner 7, which had returned startling pictures of the south polar region – had chanced to pass over the most lunar-like parts of the surface and returned pictures of what we now believe to be primitive, cratered areas. Given a difference of as little as six hours in the arrival times of any of these earlier spacecraft (each of which had spent many months in transit) an entirely different view of Mars would have resulted. It was almost as if the spacecraft from some civilisation had flown by Earth and chanced to return pictures only of its oceans."

There was initially concern that Mariner 9's systems might not operate long enough to complete the delayed mapping, but it was able to undertake an *extended* programme for almost half of Mars's seasonal cycle. When the propellant for its attitude control system finally ran out on 27 October 1972, it had returned a total of 7,329 images. NASA's administrator, J.C. Fletcher, congratulated the JPL team for making a "scientific reconnaissance of exceptional quality". With a global map at a resolution of about 1 kilometre per pixel, geologists were able to use the principle of superposition to determine the surface morphology. The dark streaks that extend for hundreds of kilometres downwind of craters proved that the large-scale albedo variations were the result of dust blown by prevailing winds. In addition, detailed views of dark patches inside craters showed fields of dunes. The greatest insight, however, was that the planet appeared to have undergone a change of climate. Life may have developed in the past, and have adapted to the change in the environment. As Sagan pointed out, "the only way to settle that argument is to land on the surface and look".

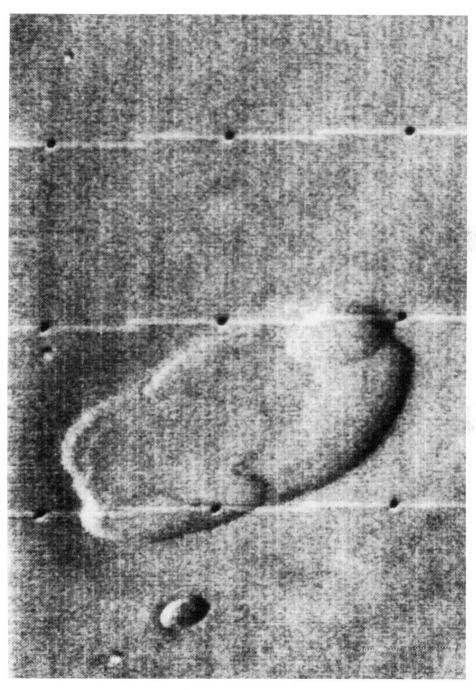

This picture taken by Mariner 9 on 27 November 1971 as the dust storm began to clear showed the overlapping calderas at the summit of the giant volcano Olympus Mons.

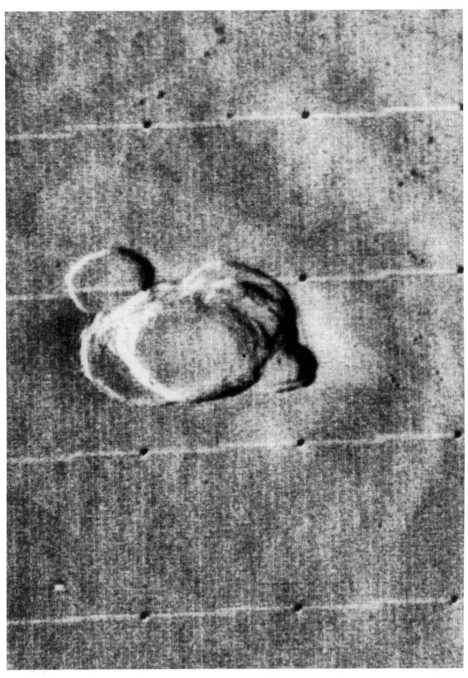

As the dust storm cleared, Mariner 9 returned this view of the northernmost of the line of three 'dark spots' that it had discerned on approaching the planet, showing the crater complex at the summit of the volcano Ascraeus Mons.

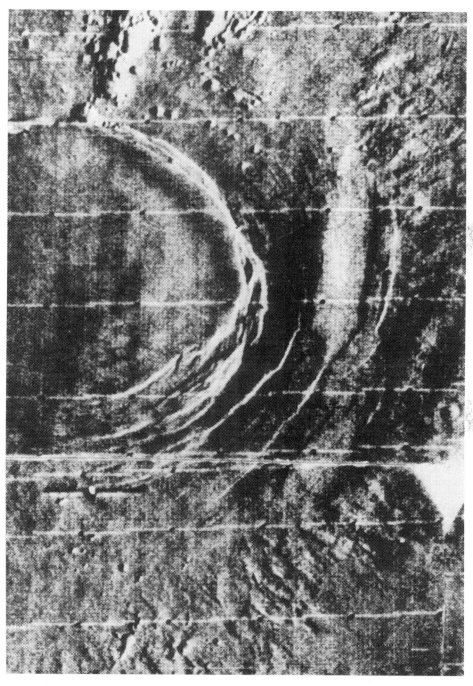

When the southernmost of the 'dark spots' was revealed by Mariner 9 to be another volcano with a summit caldera, Nodus Gordii was renamed Arsia Mons.

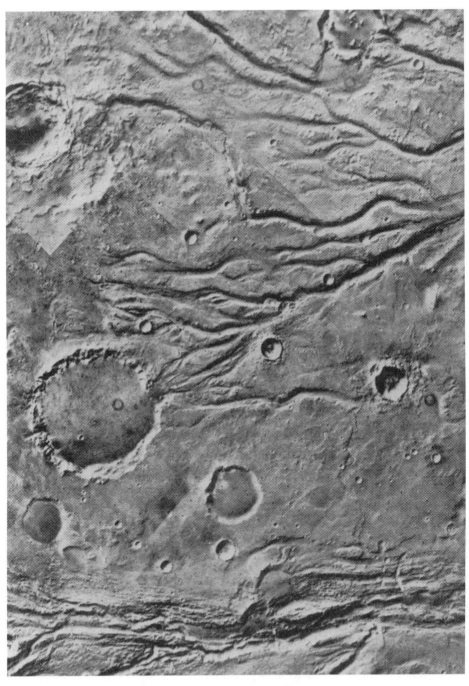

A Mariner 9 mosaic of the tributary systems draining eastwards off Lunae Planum. The small dark halos (dubbed the cheeri-oh) are due to dust on the camera's lens.

This Mariner 9 mosiac shows an extensive dendritic valley network in the southern highlands. Note the evidence of larger-scale erosion, such as the truncated form of the large crater (lower left).

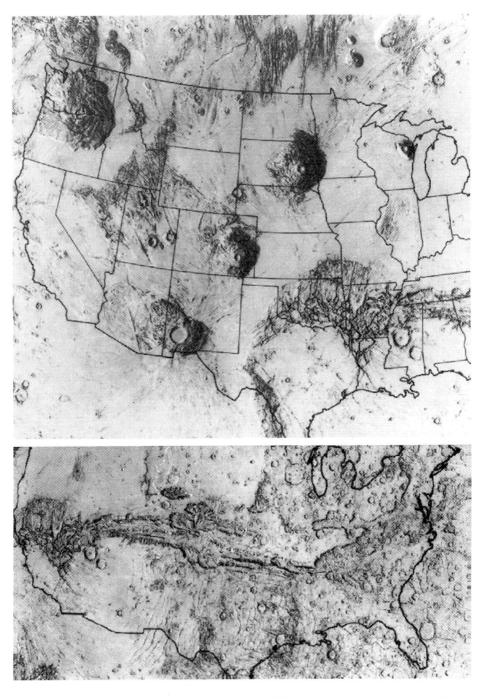

To provide a sense of the scale of the volcanoes of Tharsis (top) and the canyons of the Valles Marineris, they are superimposed on outlines of the United States.

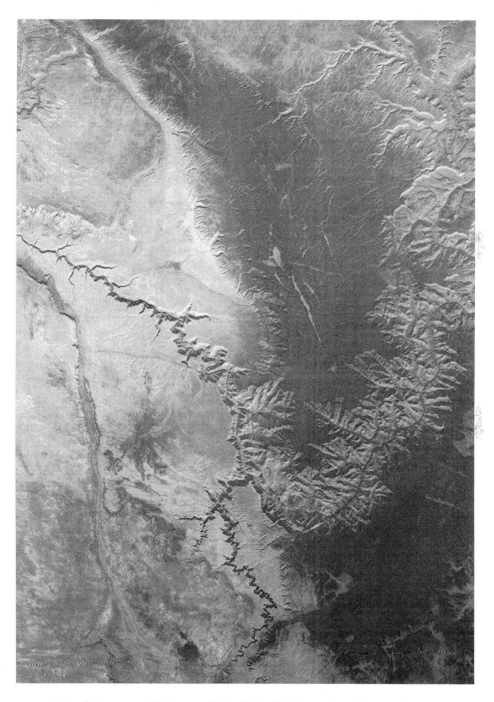

A view from space of drainage off the Colorado Plateau into the Grand Canyon.

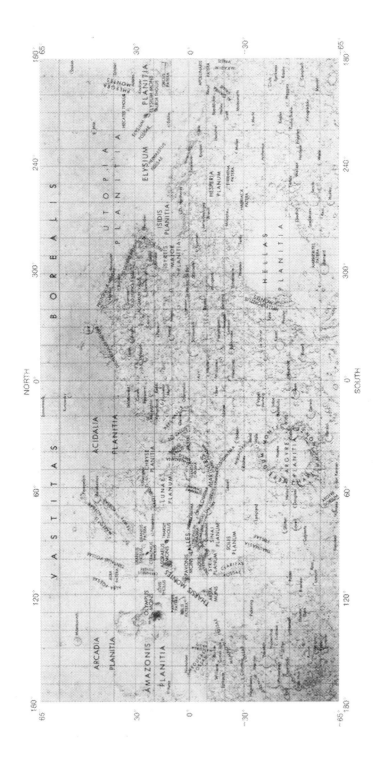

A map of Mars (with north at the top and a revised nomenclature) issued by the International Astronomical Union in 1973.

3

The Vikings

HOW TO SEARCH FOR LIFE

In 1871, 12 years after publishing *On the Origin of Species by Natural Selection* in which he expounded his theory of evolution, Charles Darwin speculated in a letter to his friend Joseph Hooker that life developed spontaneously in "some warm little pond". However, this was not a popular view. In 1924 the Soviet biochemist A.I. Oparin published *The Origin of Life*, in which he said that the synthesis of large, complex organic compounds could not have occurred in an oxygen-rich environment. He speculated that because Earth had condensed from a nebula that was rich in carbon, hydrogen, oxygen and nitrogen – collectively referred to as CHON – its primitive atmosphere was composed mostly of the hydrogen-bearing gases methane, ammonia and water vapour – a reducing environment in which such synthesis *could* have occurred. In 1929 the English biologist J.B.S Haldane noted that photodissociation by sunlight would have rapidly transformed a hydrogenated atmosphere into one dominated by carbon dioxide. Both scientists thought that a 'primordial soup' of complex organic molecules developed in the ocean and spontaneously gave rise to life. In working for his doctorate under the supervision of H.C. Urey at the University of Chicago, S.L. Miller set out to test this hypothesis by exciting a mix of methane, ammonia and water vapour using an electrical spark, and within a week had created a brown sludge containing two of the amino acids that are the building blocks of proteins. The results of this experiment, published in 1953, marked a milestone in the study of the origin of life. Later tests showed that there was no 'correct' atmosphere – it had only to include molecules that contained the elements that were most common in the solar nebula. Nor was the spark unique, as irradiating the mixture using an ultraviolet lamp gave similar results. The inference was that life was the *inevitable result of chemical evolution*. In a follow-up paper in 1959, Miller and Urey wrote: "Surely one of the most marvelous feats of the twentieth century would be the firm proof that life exists on another planet." They were thinking in particular of Venus and Mars.

The assumptions
When NASA asked the Space Science Board of the National Academy of Sciences to assist in developing a strategy to determine whether life exists on Mars, Joshua

Lederberg of Stanford University hosted a summer study in 1964 to investigate the issues. At that time, it was widely believed that the seasonal variations of the dark areas on the planet were due to vegetation, which was encouraging. In March 1965 the draft report, *Biology and the Exploration of Mars*, said it would be reasonable to assume that life originated independently on the planet. However, it pointed out, whereas if there were plants there would certainly be microbes, it was possible that there were *only* microbes, hence any test for life should be aimed at microbial life. Furthermore, the report pointed out, "We have reconciled ourselves to the fact that early missions should assume an Earth-like carbon–water type of biochemistry as the most likely basis of any Martian life." In view of the manner in which cells function, one methodology was to seek evidence of cellular reproduction, but this was a discontinuous process, the rate of which varied greatly from species to species, and even in different conditions for a given species, which would make employing it as a test very difficult in the context of an exotic environment. As an *ongoing* process that could be measured in several ways – for example, by changes in acidity or the evolution of gases – metabolism was more readily testable and was more likely to produce a definitive result. The report urged a *multifaceted* test because, "no single criterion is fully satisfactory, especially in the interpretation of negative results". It recommended "a substantial" effort to exploit the favourable launch windows in

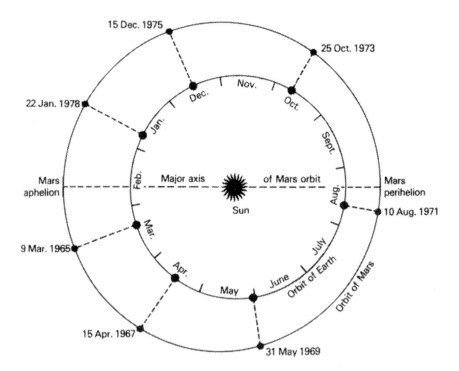

The elliptical character of Mars's orbit made the 1971 opposition particularly favourable for sending a heavy spacecraft.

1969, 1971 and 1973, with the first landing "in 1971, if possible" and certainly "no later than 1973". The results from Mariner 4's flyby of Mars several months after the draft report was circulated significantly diminished the prospect of there being life, but did not invalidate the strategy for seeking it.

Go ahead

In December 1965 NASA decided to send two upgraded Mariners on Mars flybys in 1969 and to skip the 1971 window. In 1967 it formed the Lunar and Planetary Mission Board to advise on the scientific objectives of missions, and in October 1968 this recommended making the landing in 1973. The following month funding was assigned to place spacecraft into orbit around Mars in 1971 to map the planet and identify likely landing sites. On 4 December NASA named the landing mission 'Viking', and appointed James Martin of the Langley Research Center in Virginia to manage it. In January 1970 financial constraints obliged a postponement to the less favorable 1975 launch window.

THE VEHICLES

The mission architecture called for the lander to be ferried into Martian orbit by a mothership, then released to make the landing. Like Mariner 9, the orbiter was built on an octagonal framework, but with larger solar panels to provide 620 watts in Mars orbit supplemented by a pair of 26-cell nickel–cadmium batteries. It had a rocket engine mounted on a gimbal that burned monomethyl hydrazine in nitrogen tetroxide, delivering a thrust of 136 kilograms. This was to execute the mid-course corrections during the interplanetary cruise, the orbital insertion burn, and such orbital trims as required to adjust the inclination plane of the orbit and maintain the desired ground track. The attitude control system utilised a star tracker and a Sun sensor, and squirted gaseous nitrogen to maintain orientation with respect to these references. A similar system had been used by Mariner 9, but auxiliary piping had been added to enable propellant to be drawn from the main propulsion system after the nitrogen was expended. The orbiter had a high-gain antenna with a 1.5-metre-diameter motorised dish for rapid downloading of science data. It received S-Band commands, and the 20-watt transmitter could send S-Band data at between 2 and 16 kilobits per second. A rod-like omnidirectional antenna on the side of the vehicle would maintain low-gain communications when the high-gain antenna was unable to point Earthward. The data storage system had twin 8-track digital tape recorders, each with a capacity of 640 million bits. One track on each recorder was for engineering data, and the others were for science data. The vehicle was run by a redundant computer system, each unit of which had a memory capacity of 4,096 words. Part of this capacity was preloaded and write-protected, and the remainder was available for storing uplinked commands.

The scan platform had three bore-sighted instruments. The imaging system had two television cameras set side-by-side. Each had a Cassegrain telescope with a focal length of 475 millimetres, giving a field of view of 1.54 by 1.69 degrees that would

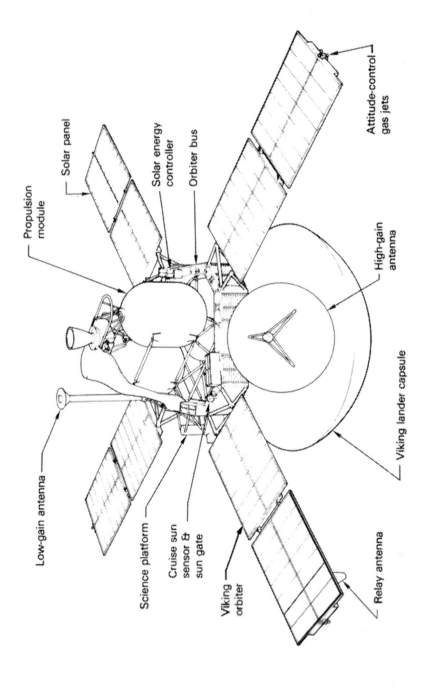

A schematic of the structure of the Mariner-derived Viking orbiter with its lander housed in the biological isolation capsule.

cover a surface area of 40 by 44 kilometres at the planned 1,500-kilometre periapsis. The system used the same technology as the 1971 orbiter, in which an electron beam read the electric charges off a vidicon tube on which the image was made. Each image comprised 1,056 lines of 1,182 pixels with a 7-bit grayscale. As the acquisition of a frame and its transfer to tape would take 8.96 seconds, alternating the cameras enabled a frame to be taken every 4.48 seconds. This alternation, combined with the spacecraft's motion along its orbit, combined to record a swath running along the ground track. The imaging team was headed by Michael Carr of the US Geological Survey at Menlo Park in California. To the left of the cameras was a spectrometer to measure the diffuse reflection of sunlight at a wavelength of 1.38 micrometres in order to measure how much water was present as vapour in the atmosphere. At periapsis, this observed a 'footprint' some 3 by 20 kilometres, which a nodding mirror split into 15 rectangles for sampling, taking a full set of readings every 4.48 seconds, which was the same rate as the imaging. The instrument had a sensitivity of better than 1 precipitable micron, which is how the vapour content of an atmosphere is measured – the Earth's being between 1 and 5 centimetres. Because Mars's atmosphere is so tenuous, most of the water vapour is concentrated in the lowest kilometre and the readings could be interpreted in terms of the terrain elevation. To the right of the scan platform was the infrared thermal mapper that comprised four small telescopes, each with seven antimony–bismuth detectors to distinguish infrared reflection from infrared emission in the range 0.3 to 24 micrometres. During a 57-second interval it measured a series of spots, each 8 kilometres across at periapsis, arranged in an inverted chevron, after which a mirror flipped the field of view to space in order to set a 'zero' point; the entire cycle taking 1.25 minutes. As large rocks retain heat for longer than fine material, diurnal temperature gradients would yield insight into whether the surface was sandy or rocky, which would be an important factor in determining the best site at which to set down the lander.

The lander

On being released, the lander was to make an autonomous landing and operate on the surface for a minimum of 90 sols. It was a six-sided aluminium–titanium box 3 metres wide and 0.5 metre deep, with three sides about 1 metre in length, and the others – on which its legs were installed – of half that length. The interior, which was insulated against heat loss, housed the computer, data storage system, tape recorder, radios, batteries, control systems, biology package, X-ray fluorescence spectrometer and molecular analysis instrument. All external apparatus was painted white to reflect insolation for thermal regulation. The primary command receiver was a low-gain antenna. A 1-metre parabolic dish on an articulated mast was electrically driven to track Earth as Mars rotated for direct S-Band transmission at 1 kilobit per second. In addition, a UHF transmitter could operate at 1, 10 or 30 watts, and send up to 16 kilobits per second to the orbiter. The activities of the lander were managed by its guidance, control and sequencing computer, which comprised two redundant units, each with a 25-bit 18,432-word memory. In the event that the lander was unable to receive commands, it was to undertake a 60-sol sequence (two-thirds of the nominal duration of the primary mission) that would be uploaded shortly prior to

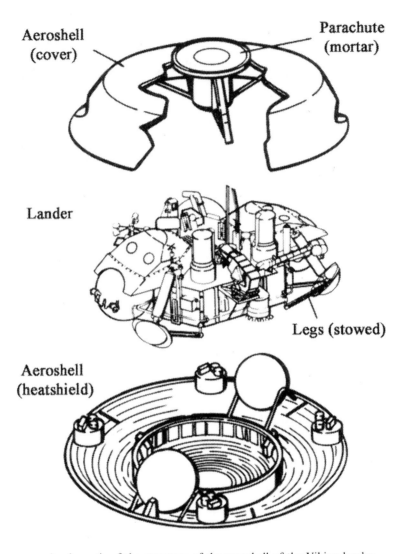

Aeroshell (cover)

Parachute (mortar)

Lander

Legs (stowed)

Aeroshell (heatshield)

A schematic of the structure of the aeroshell of the Viking lander.

separation. If all went well, the computer would be updated in 3-sol cycles with a list of activities to perform. At times when the lander was unable to transmit to either the orbiter or Earth, the data acquisition and processing unit would store engineering and science data on a recorder that had 200 metres of four-track tape with a total capacity of 40 million bits. The lander would be able to uplink by UHF for at most 32 minutes each time that the orbiter made a periapsis passage over the site, with the data either being relayed to Earth in real-time or being stored on tape. The lander had two radioisotope thermal generators with a combined output of 70 watts. When more power was required, it was able to draw on four rechargeable

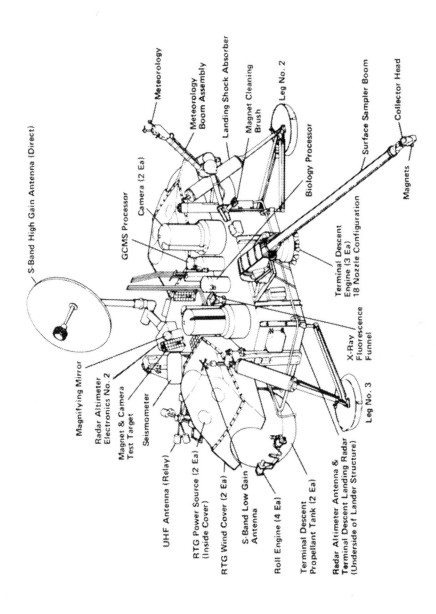

A schematic of the structure and instruments of the Viking lander.

A Titan-III–Centaur lifts off with Viking 1 on 20 August 1975.

nickel–cadmium batteries. A treaty signed in 1967 required that spacecraft sent to Mars be sterilised to preclude contaminating the planet with terrestrial microbes – and avoid the nightmare scenario of life-detecting instruments detecting pollution from their own vehicle! It was decided to seal the lander in its bioshield and bake it for 24 hours at 113°C, which trials indicated would reduce the probability of a single microbe hitching a ride to the Martian surface to less than 1 in 10,000. The challenge for the engineers, of course, was to design the systems to withstand this treatment. The medium for the tape recorder, for example, was made of a phosphor–bronze base coated with nickel–cobalt. The aeroshell housed attitude control thrusters, rockets for the de-orbit burn, and experiments to investigate the upper atmosphere during the entry phase, and the outer surface of the lower part had cork and glass beads embedded in silicone as a thermal shield.

ON MARS

Viking 1 was launched on 20 August 1975 and went into orbit around Mars on 19 June 1976. The operational constraint on inspecting the nominal landing sites was that although the longitude of periapsis could be adjusted, there would be insufficient propellant to alter the latitude once this had been set. Since water vapour would be more abundant in the summer hemisphere, with its retreating polar cap, the targets for both landers were in the northern hemisphere – at 22 degrees north in the case of Viking 1. The prime site was on the sedimentary plain of Chryse Planitia at 35 degrees (Martian longitudes are west), and the backup was in the Elysium volcanic province at 255 degrees. Selecting a site was not just a matter of choosing a point, since the intrinsic uncertainties of the entry procedure meant that the target was an elliptical 'footprint' extending 120 kilometres in the direction of travel and 25 kilometres to either side of that track. If the lander was aimed at the centre of the ellipse, there was expected to be a 99 per cent chance of its reaching the surface within this boundary. There was an imperative to 'certify' a site, because the plan called for Viking 1 to land on 4 July in order to mark the American Bicentennial. Because the Mariner 9 imagery had generally been able to resolve surface features no smaller than about 1 kilometre across, the candidate sites had been examined by a radio telescope acting as a radar, by which a strong reflection indicated a smooth surface, and a weak signal indicated a rough surface. Although the radar operating at a wavelength of 13 centimetres offered insight into the nature of the terrain on the 1-metre scale, the fact that the reflection was averaged over a wide area meant that a strong signal did not indicate an *absence* of boulders, merely that there seemed to be *few*. Since the belly of the lander had only 22 centimetres of clearance above the ground, it would still be wrecked if it were unfortunate enough to land on the only boulder marring an otherwise smooth plain. The orbiter took its first picture of Chryse on 22 June, and the 200-metre resolution revealed a profusion of small craters, which was bad news as impacts throw out blankets of rocky ejecta. After further imagery showed craters, channels and cliff-edged mesas all across the prime site, Martin announced on 27 June that the landing would be postponed. The pressure on the site selection team, led by Hal Masursky of

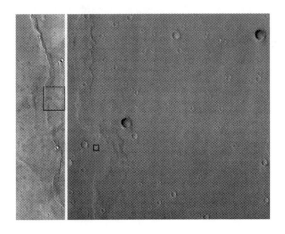

An orbital view of the Viking 1 landing site. (see page 185 for larger context)

the US Geological Survey, was intense, because the landing had to be attempted before 22 July, when Viking 2 would have to commit to a latitude for the periapsis of its initial orbit. A trim burn enabled Viking 1 to examine a site 250 kilometres northwest of the nominal target, since the terrain appeared to become smoother in that direction, but this was also too rough. Attention switched to a site 580 kilometres further west, where there were fewer fresh-looking craters, and on 13 July Martin decided to attempt to land at 22.5 degrees north and 47.4 degrees longitude on 20 July. If it failed, there would be two days in which to decide the latitude of periapsis for Viking 2.

Touchdown!
On 17 July the flight controllers activated the Viking 1 lander for its pre-separation checkout. The power from the radioisotope thermal generators was brought on line the next day, in preparation for severing the umbilical to the mothership. At 23:00 local time on 19 July Mission Director Tom Young in the Space Flight Operations Facility in Building 264 of the JPL campus gave the 'GO!' command – which was sent by the Deep Space Communications Complex in Australia and reached Mars 18 minutes later. Once the spacecraft was in the requisite attitude, three explosive bolts fired to release the lander. When the signal verifying that the lander was free and stable was received at 01:51:15 on 20 July, the controllers cheered. The lander would operate autonomously. If anything went awry, its masters were powerless to intervene; they were merely spectators. The separation had occurred at an altitude of 5,000 kilometres as the spacecraft pursued its elliptical orbit. At 01:58:16 the lander began the de-orbit burn. The 4-kilobit-per-second telemetry showed that the thrust was nominal. The burn ended at 02:20:32, having slowed the vehicle by 156 metres per second, just right to enable it to contact the atmosphere on a trajectory depressed about 16 degrees below horizontal at a point that, if things went to plan, would result in a landing within the target ellipse. As it entered the atmosphere at 05:03, sensors on the aeroshell began to measure the composition, temperature and pressure. At an altitude of 27 kilometres, the dynamic load reached a peak of 8.4 g, at which time the vehicle had slowed to a speed of 2 kilometres per second.

Despite the early hour, the von Kármán Auditorium was packed. In addition to 400 journalists from around the world, there were 1,800 invited guests watching a closed-circuit television view showing the control room, with Albert Hibbs, one of the mission planners, providing the commentary.

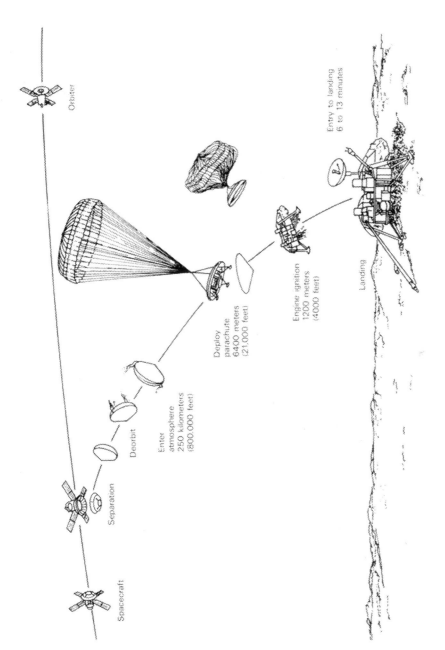

A schematic of the major events for Viking entry, descent and landing.

"We're coming down," Hibbs pointed out, as the telemetry showed the vehicle descending though an altitude of 24 kilometres. "It's a long period of almost flat glide to get rid of some more of the speed before the parachute comes out." As the vehicle slowed to 1 kilometre per second the load diminished to 0.8 g. At 05:10, at an altitude of about 6 kilometres, a mortar in the rear of the aeroshell deployed the 16-metre-diameter parachute and the forward shell was jettisoned. A few seconds later the three legs were deployed and locked into position. A minute after that, at an altitude of 1,400 metres, having slowed the descent to 54 metres per second, the chute was released and three throttleable rocket engines on the sides of the lander – the sides not occupied by legs – were ignited for the terminal phase. An inertial reference and a radar altimeter enabled the computer to control the rate of descent by varying the thrust. One requirement of the design of the engines was to minimise the disturbance to the surficial material, in part because the biologists did not want the exhaust gases to contaminate the soil that was to be tested for life. To disperse the plume, each engine had 18 small nozzles that pointed in a range of downward directions. The rockets were to slow the rate of descent to 2 metres per second for contact. As the system would be shut down by sensors in the base of the footpads, the first leg to make contact with something solid would terminate the controlled phase of the descent. Thereafter the vehicle would fall under gravity – hopefully in a horizontal configuration, and onto a flat surface. On receiving the contact signal the computer also switched the telemetry from 4 to 16 kilobits per second, which was noted by Hibbs: "We have touchdown!" The controllers began cheering, shaking hands and embracing. It was a historic moment. It was not merely a robot that had landed, but so too, in a sense, had humankind and, as Carl Sagan had pointed out, "You can only land on Mars for the *first time* once."

Surface view
It was late afternoon on Chryse Planitia. In fact, the landing site was beyond the limb as seen from Earth, and the lander's signal was being relayed by the orbiter as it passed overhead and dipped towards the horizon. As the next communications relay would not occur until the orbiter made its next periapsis passage, the lander's first assignment was to take pictures of the site as a contingency against something untoward occurring during the Martian night. The Itek Corporation's imaging system had two cameras mounted on top of the lander, spaced 1 metre apart for stereoscopic viewing. As the lander was initially to have transmitted its imagery directly in real-time, the scan rate of the novel cameras had been matched to the transmission rate. When the tape recorder was later added, the design of the camera was already 'frozen'. Each camera had an upward-aimed photodiode at the base of its cylinder – in fact, 12 diodes were available: one for monochrome panoramas; a red, green and blue for subsequent composition into colour; three infrared; four located at different focal points to facilitate high-resolution monochrome imagery; and one of low sensitivity for if the Sun was in the field of view. The optics were above the photodiodes, and topped by a mirror that rotated to provide a vertical sweep of 512 pixels, each a 6-bit word. The cylinder rotated axially after each sweep to extend the image horizontally, line by line. As the cameras could rotate almost 360

The first picture returned by the Viking 1 lander.

degrees they could view across the top of the deck as well as outward and, between them, take panoramic views of the site. However, doing so took a long time. In fact, when a test version of the camera was set up to take a panoramic picture of a large group of project personnel, those in shot at any given time had to stand very still, and it took so long to complete the scan that by the time the line of sight reached the other end of the group the people at the start had long-since departed!

A few seconds after landing, one of the cameras began to take the first of two monochome images. These were taped by the orbiter which, on re-emerging from behind Mars, relayed them to Earth. The first picture started to arrive at 05:47. Tim Mutch of Brown University, Rhode Island, the imaging team leader, hunched over the screen to be the first to see what Mars looked like from ground level. The first few lines showed streaks resulting from the slow-scanning camera catching dust as it settled around the craft, but once this had settled the clarity was startling. As the image built up, extending to the right, a small rock appeared sitting on a level surface of fine-grained material, and then some other rocks interspersed with pebbles. There were signs of erosion by windblown dust, and some rocks had tiny holes resembling vesicles. "It's a geologist's delight!" exclaimed Mutch. Several minutes into the scan, one of the 0.3-metre-diameter circular footpads appeared. It was resting cleanly on the surface, undamaged, and with some dust on its concave upper surface. "We always knew it was going to be good," enthused Mutch as the scan finished. It was just as if he was there himself, looking down at his feet. "It's incredible." Over the next 30 minutes a panorama built up showing a view out to the horizon, some 3 kilometres away, the line of which showed that the lander had come to rest on level ground. The terrain was strewn with rocks, some rather large, and there were several sand dunes. The geologists noted the striking similarity to the high desert of the American Southwest. As the team walked onto the stage for the post-landing press conference they were greeted by sustained applause.

On sol 1 (21 July) the first colour panorama was taken, but although there were colour charts on the lander, the picture had been rushed out without being properly calibrated and the sky was blue. A colour-balanced version released the following day revealed the sky to be *salmon pink*. In Earth's atmosphere, Rayleigh scattering by air molecules diffuses blue sunlight and makes the sky blue. The much thinner

Viking 1's first panoramic view revealed that it had landed very close to a boulder about 3 metres wide.

Martian atmosphere did not scatter blue very efficiently, but (as occurs on Earth at sunset when the air is dusty) the dust motes scattered the red end of the spectrum.

The surface sampler

The lander had an arm with which to retrieve soil for the experiments. The boom comprised a pair of thin ribbons of stainless steel welded together along the edges, which stiffened (in the manner of a steel tape measure) as it was unrolled, opening out to form a rigid tube. It had a reach of some 3 metres and could swivel across a horizontal arc of 300 degrees, elevate 35 degrees and dip 50 degrees. At the end of the boom was the collector head. To retrieve a sample, the lid would be raised, the boom driven into the topsoil, the lid closed, the boom retracted, the head rotated to dump the sample onto the lid, and then vibrated to encourage fines to pass through the small holes into the appropriate sample inlet. On sol 2 the arm rotated to enable the protective shroud on its collector head to be ejected. In doing so the arm was to extend about 30 centimetres and then return to its stowed position, but it jammed. The problem was soon found to be the locking pin in the mechanism that latched the shroud; this had not fallen out because the arm had not extended sufficiently far. As lander operations were planned in three-day uplink cycles, instructions to extend the arm 35 centimetres to shake loose the pin were added to the sequence to be executed on sol 5, and the pictures taken in support of this effort showed the pin on the ground.

The soil

The magnetic properties experiment was to test the hypothesis that the surface was reddish due to limonite, an oxide of iron. In fact, iron oxidises in a number of ways, and not all oxidation products are magnetic. The lander carried magnets – one on the colour calibration chart and two on the rear of the arm's backhoe. The magnets had a characteristic bull's-eye pattern, and any surficial material attracted to the magnets should duplicate this pattern. The plan was to identify the forms of oxidised iron from the colour of the material (once isolated from the non-magnetic material in the soil) and the strength of its magnetic attraction. As the arm began to investigate

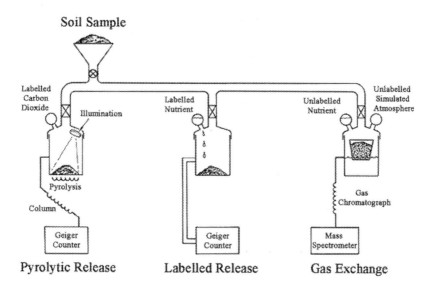

A functional schematic of the three biology experiments.

the surface, Robert Hargraves of Princeton University, who headed this team, saw that both the strong and the weak magnets on the backhoe were totally covered by tiny particles in the bull's-eye pattern, and concluded that it was one of four types: (1) magnetite, (2) maghemite, (3) iron or (4) sulphide pyrite. The sample of soil provided by the arm to the X-ray fluorescence spectrometer on sol 8 was irradiated to cause the nuclei of the constituent atoms to emit at characteristic frequencies, and thereby give rise to a series of electrical pulses in the detectors in proportion to the energy of the X-rays. The pulses were counted over a succession of intervals. Since X-ray analysis is a statistical process, the more data the greater the accuracy. This experiment gave the chemical composition of the soil in terms of its *elemental* abundances, from which the team headed by Priestley Toulmin of the US Geological Survey in Reston, Virginia, set out to infer the likely *mineralogical* abundances. Their analysis ruled out the pyrite, and native iron was unlikely in such an oxidising environment. The two remaining candidates represented different stages of iron oxidation. One surprise was that the iron oxide content was relatively low – implying that it was a thin coating 'staining' things. The main elements were iron, calcium, silicon, titanium and aluminium. The closest terrestrial analogue to the sample was basalt from Molokai in the Hawaiian Islands, indicating that the dust was weathered basalt.

The biology experiments

The Gas Exchange experiment developed by Vance Oyama of the Ames Research Center in California was initiated on 29 July. It assumed that Martian life would resemble terrestrial life, and that if the planet underwent episodic climate variations then its microbes might go dormant during the long, cold, arid times, awaiting a

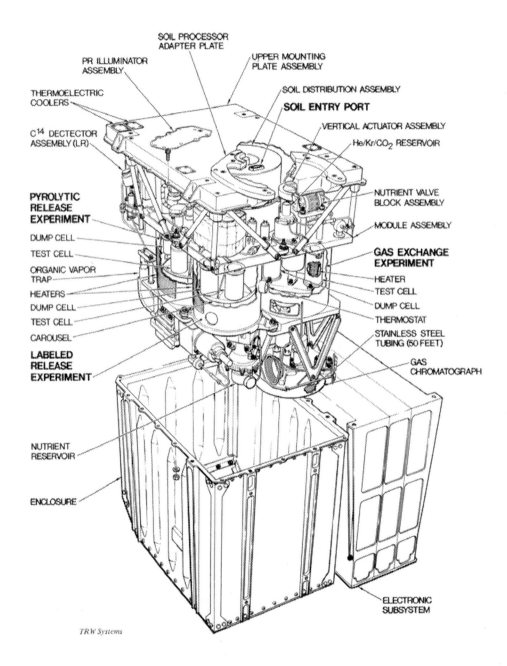

A schematic showing the integrated hardware of the Viking biology package.

resumption of more benign conditions – which the experiment aimed to offer. The objective was to determine whether metabolism caused changes in the *composition* of the gases in the test chamber. To promote metabolism, the sample was to be provided with an aqueous solution of nutrient dubbed 'chicken soup' that had almost everything a terrestrial microbe might consume: amino acids, purines, pyrimidines, organic acids, vitamins and minerals. Because water vapour was not stable at the Martian surface, the pressure in the chamber had to be significantly greater than ambient in order to prevent the nutrient breaking down. The experiment was to proceed in two stages. First, a mix of carbon dioxide and krypton was added to the chamber and a mist of nutrient introduced to expose the sample to water vapour. Two hours later, a sample of the gases in the chamber was sent to the Gas Chromatograph and Mass Spectrometer to set the benchmark for comparison with analyses at various stages of the incubation seeking metabolic products such as hydrogen, oxygen, nitrogen, carbon dioxide and methane. This initial measurement showed a surprisingly large peak for oxygen – fully 15 times as much as could be accounted for by adding up the known sources in the atmosphere and the added gases. Was this the result of an orgy of metabolism as the nutrient awakened dormant microbes? The onus on the scientists was not to proclaim life as soon as they saw a response that *could* imply biology, but to err in favour of chemistry until they saw something that *could only* be explained by biology. For the so-called 'humid mode' of the test, which was to last a week, the sample was suspended in a porous cup above the nutrient, and not allowed to come into direct contact with the solution. By 1 August, the release of oxygen had slowed considerably. The rapid release of oxygen, slowly tailing off, suggested that there was an intense but brief inorganic reaction between the soil and the water in the nutrient. Free hydroxyl ions from ultraviolet dissociation of water vapour near the surface would build up peroxy compounds in the chilly, dry soil. Peroxides, superoxides and ozonides are all strong oxidising agents, but in the presence of significant water vapour they would rapidly break down into water and gaseous oxygen. Was this why so much oxygen was evolved by the sample when presented with a mist of water vapour? On 5 August the experiment advanced to its 'wet mode', with the injection of nutrient directly into the sample. This caused one-third of the carbon dioxide in the chamber to be dissolved by either the sample or the water, then, as the six-month incubation proceeded, it gradually returned to its initial level and activity ceased. The *uptake* of carbon dioxide was explicable as water causing the peroxy compounds in the soil to draw in carbon dioxide to create metal oxides or hydroxides, and the later slow *release* of this gas was explained by iron oxides in the soil reacting with nutrients dissolved in the water and liberating it. Meanwhile, oxygen was reclaimed instead of being released, with the amount of oxygen taken up matching the ascorbic acid (vitamin C) in the nutrient. Evidently, the Martian soil was extremely reactive.

The Labelled Release experiment, developed by Gilbert Levin of Biospherics Incorporated, and initiated on 30 July, made fewer assumptions concerning the biochemistry of Martian life. On the assumption that life would be adapted to its environment, the nutrient was a weak 'broth' of glycine and alanine amino acids, and formic, glycolic and lactic acids in the form of salts in distilled water, at least one of which ingredients, it was thought, could probably be metabolised by an alien carbon-based lifeform. The premise was that if microbes consumed the nutrient, their metabolism would produce gases such as carbon dioxide or methane that would be detected by labelled carbon-14 using a Geiger counter. In contrast to the Gas Exchange experiment, however, this experiment did not determine *which* gases were present. A mist of nutrient was introduced to the chamber to moisten the soil, and helium was injected to maintain sufficient pressure to prevent the nutrient breaking down. A rapid rise to 10,000 counts per minute indicated that a large amount of gas was evolved as soon as the nutrient was added, but by 2 August it was evident that there was no exponential increase to suggest growth. This result did not match *either* the predicted biological or chemical responses. Once it was apparent that the counts had levelled off, meaning that the release of radioactive gases had ceased, a second injection of nutrient was made. If the initial evolution of gas had been due to microbial metabolism of the nutrient, there should have been a second rise, but the rate rapidly fell to 8,000 counts per minute and levelled off, indicating (a) that the micobes had expired after their initial feast, or (b) that the reaction was inorganic. In view of the evidence for peroxy compounds in the sample, hydrogen peroxide could have oxidised the formic acid in the nutrient to carbon dioxide and water. The amount of radioactive carbon dioxide was only slightly less than would have been produced if all of the formic acid had been consumed in this manner. If the oxygen emitted in the Gas Exchange experiment was due to the dissociation of peroxides in the soil by the water vapour of the nutrient mist, then the water produced in the Labelled Release experiment by the break-down of formic acid should have decomposed *all* of the peroxy compounds at the first injection of nutrient, and the second injection should have produced no additional radioactive gas. The fact that the count fell when the nutrient was topped up implied that some of the carbon-14 was being reclaimed, probably by carbon dioxide being absorbed by the water in the nutrient.

The Pyrolytic Release experiment developed by Norman Horowitz of Caltech was initiated on 28 July. It assumed that any Martian microbes would be adapted to the environment, but also require to 'fix' carbon from the atmosphere. It was to seek evidence of *synthesis* of organic matter in conditions as close as possible to the local environment. The soil was sealed into the chamber, and the air evacuated and replaced by a representative mixture of carbon monoxide and carbon dioxide labelled with carbon-14. During incubation, the chamber was illuminated by a xenon arc that simulated sunlight at the Martian surface, minus the ultraviolet. After five days, the lamp was to be turned off and the gases flushed out and analysed by a Geiger counter to produce the 'first peak' that measured the carbon-14 left in the air. To determine whether microbes had assimilated carbon, the sample was to be heated to 640°C to pyrolyse the organic molecules and thus release the carbon-14 as carbon dioxide, which would be measured as the 'second peak'. A high second peak would support a biological interpretation, but a low one would indicate that there were few, if any, microbes (with the strength of this conclusion depending on how close this peak was to non-existent). On 7 August it was announced that the second peak was strong, but again there was ambiguity. The results argued against the peroxides that were candidates for a chemical interpretation of the Gas Exchange's response, since the Pyrolytic Release had seen a *reducing* as opposed to an *oxidising* reaction, and peroxides would have tended to decrease the size of the second peak.

Next, the Labelled Release and Pyrolytic Release experiments were repeated as 'control' tests in which, prior to incubation, the samples were heated sufficiently to kill microbes but not to inhibit most chemical reactions. If the responses seen in the initial tests were the result of biology, then they ought *not* to be repeated with these sterilised samples. Levin reported on 20 August that although the control test of the Labelled Release experiment *could* have ruled out biology, it *did not do so* – the radioactivity had rapidly risen to 2,200 counts per minute, fallen back sharply, then levelled off at 1,200. He was encouraged: "If we'd run this experiment in the parking lot at JPL [and seen these two curves] we'd have concluded that life is present in the sample." However, if it was a chemical reaction it was one that was disabled by heat. "We've significantly narrowed the range of possible chemical explanations." If the initial carbon assimilation of the Pyrolytic Release had been due to biology, the sterile sample ought to have been negative. In fact, there was assimilation, albeit at a much reduced level, which argued for a chemical reaction involving several reactants, only some of which had been inhibited by pre-heating. When the Gas Exchange experiment had finished its six-month incubation, it was emptied and reloaded with fresh soil, which was sterilised before rehumidification. The fact that half of the initial amount of oxygen was released indicated that there was a non-biological reaction. Because the pre-heating would have dissociated hydrogen peroxide, this suggested that the reaction was produced by a more thermally stable superoxide.

The first analysis by the Gas Chromatograph and Mass Spectrometer was done on 6 August. To start, the sample was heated to 200°C in order to drive most of the water out of the hydrated minerals, but surprisingly little water was released. Next, the sample was heated to the maximum 500°C to volatise organic molecules. The astonishing fact that there were *no organics* was reported on 13 August. However, the analysis had been complicated by the delayed release of water. Another sample was tested on 21 August to gain further insight. A significant amount of water was liberated at 350°C, but despite the improved sensitivity in the second stage, the organics present, if any, were below the 10–100 parts per billion detection limit. However, as team leader Klaus Biemann of the Massachusetts Institute of Technology noted, there would have to be at least 1 million microbes in the sample for their organics to be detected at that sensitivity. A typical temperate sample of terrestrial soil can contain hundreds of millions of bacteria per cubic centimetre. If only the living cells were present for analysis, then 1 million bacteria would have been far too few for the instrument to detect. In terrestrial soil, the amount of dead organic matter often outweighs the living material by a factor of 10,000, and if Martian microbes were the same they would probably have been able to be detected by the organic wastes and dead cells they produced. But if they recycled their wastes and solar ultraviolet destroyed the dead cells, microbes may well have been present to produce the reactions reported by the biology package without their being detected by the spectrometer. Levin later processed samples from Antarctica and confirmed that soil that the Gas Chromatograph and Mass Spectrometer reported to be devoid of organics had nevertheless contained sufficient bacteria to replicate the results of the Labelled Release experiment on Mars. Although this hypothesis explained the conflicting results, there was no proof that it was true.

As a footnote, after Tim Mutch's untimely death in a climbing accident in the Himalayas on 7 October 1980, NASA created a plaque for the first humans to visit Mars to affix to the Viking 1 lander, renaming it the Thomas A. Mutch Memorial Station.

A second site

On arriving on 7 August, Viking 2 entered an orbit with a periapsis at a latitude of 44 degrees north, as planned. Because a reconnaissance by Viking 1 had shown the intended landing site in Cydonia to be too rough, the newcomer's first task was to search for somewhere more suitable. If necessary, Martin was willing to postpone this second surface mission until after Mars had passed through solar conjunction in November. The inability of terrestrial radar to provide information regarding the smoothness of terrain at the target latitude was compensated by using the orbiter's infrared thermal mapper which, at periapsis, gave insight into whether the surface was blocky. After orbital adjustments to extend the search northward, a site at 47.9 degrees north and longitude 225.9 degrees on Utopia Planitia was selected, and the landing scheduled for 15:58 PDT on 3 September. The separation of the lander at 12:40 seemed nominal, but five seconds later the orbiter's main gyroscopes began to lose power. Four minutes after that the vehicle began to drift in attitude and the high-gain link deteriorated. About three minutes later, with the primary attitude control system making seemingly excessive thruster firings, control was handed to its backup and, as a precaution, the high-gain antenna was abandoned in favour of the omnidirectional antenna, which did not have the capacity to relay the lander's telemetry. On finding that it too was making excessive thruster firings, the backup attitude control system switched to a second set of gyroscopes, which resolved the issue and enabled the vehicle to stabilise itself. Lacking telemetry, the mood in the Space Flight Operations Facility was even more anxious than for the first landing. The engineers could only monitor the low-gain signal from the orbiter, awaiting an indication that the lander had switched from 4 to 16 kilobits per second, which it was to do on landing, and it was 21 seconds after the predicted moment before this switch

The first picture returned by the Viking 2 lander.

Utopia Planitia at the Viking 2 landing site.

over occurred. As the lander worked through its initial post-landing activities, the engineers concluded that one of the explosive bolts of the separation system must have damaged the orbiter. Fortunately, the orbiter had been able to recover in time to record the pictures from the lander and these were relayed once the high-gain link was re-established.

Contrary to expectation, Utopia was strewn with rocks, one of which was so close to the documented footpad that it may well have been nudged aside during the landing. The chemical analysis of the soil proved it to be very similar to that on Chryse Planitia, on the opposite side of the planet. As regards the biology package, it placed tighter constraints on putative inorganic chemistry but did not resolve the debate.

WHAT WAS CONCLUDED?

Gilbert Levin opined, "The accretion of evidence has been more compatible with biology than with chemistry – each new test result has made it more difficult to come up with a chemical explanation, but each new result has continued to allow for biology." All other things being equal, he noted that if a terrestrial sample had given the observed results, "we'd unhesitatingly have described [it] as biological". Vance Oyama was sceptical: "There was no *need* to invoke biological processes." Norman Horowitz agreed, but admitted it was "impossible to prove that any of the reactions ... were *not* biological in origin". Prior to Viking, nobody knew whether there was life on Mars; and, sadly, nobody knew afterwards either!

Harold Klein of Ames, leader of the biology team, later recommended that the assumption that Martian microbes would be similar to terrestrial ones should be dismissed, and that scientists ought to consider whether the Viking data suggested any clues as to "whether there might be some less obvious kind of life on Mars". In fact, even as the Vikings were seeking carbon-synthesising microbes, biologists on Earth were discovering the first examples of a whole new class of microbial life which, if its existence had been known when Viking was being planned, would have prompted a wider range of experiments.

4

What is life?

CELLS

It was once believed that the smallest living organisms were insects, but when the microscope was invented early in the seventeenth century it revealed a whole new realm of nature. On noticing in 1665 that cork has a distinctive structure, Robert Hooke, the curator of the Royal Society of London, introduced the word 'cell'. In Holland in 1675, while analysing a drop of water, Anton van Leeuwenhoek found organisms that were too small to be seen by the naked eye and realised that insects were not the smallest creatures after all. Further investigation revealed that these organisms were a type of animal that was able to move by flexing small *flagellae* (whips) or *cilia* (hairs). In France in 1824, R.J.H. Dutrochet proposed that living matter is composed of cells. This was confirmed for plants by M.J. Schleiden in Germany in 1838, and for animals by Theodor Schwann in Germany in 1839. In each case the cells were typically between 5 and 40 micrometres in diameter. In 1839 the Czech physiologist J.E. Purkinje named the colloidal fluid in cells *protoplasm* (first form) and the German anatomist M.J.S. Schultze later demonstrated the similarity of protoplasm in plant and animal cells, and said it was the "physical basis of life". The *spermatozoa* (animal seed) had been seen in 1677 by Johann Ham, one of van Leeuwenhoek's assistants, and in 1827 the German physiologist K.E. von Baer identified the mammalian *ovum* (egg), but it was not until the 1870s that microscopes became sufficiently powerful to reveal that the fertilisation of an egg by a sperm led to the division process that formed a multi-cellular organism. In 1845 K.T.E. Siebold discovered that the organisms that van Leeuwenhoek had observed in water droplets were "animals whose organisation is reducible to one cell", and in 1848 he defined them collectively as *protozoa* (first animals) because they appeared to be the basis of the 'animal kingdom'.

In 1676 van Leeuwenhoek discovered *germs*, but they were at the limit of the power of a microscope at that time, and it was almost 100 years later before they could be studied. While some of these entities could move, most were immobile. Because they did not have chlorophyll it was evident that they were not plants, but nor had they much in common with animals. They were therefore classified with fungi, which appeared to be a plant without chlorophyll that lived off organic matter.* In 1773 the Danish microscopist O.F. Müller distinguished two types of germs: *bacilli* (rods) and *spirilla* (spirals). When the Austrian surgeon Theodor Billroth found the even smaller *coccus* (berry), it was realised that germs were much simpler than Siebold's protozoa. In 1872 the German botanist F.J. Cohn introduced the name *bacterium* for the entire class. In 1878 Emmanuel Sedillot coined the term *microbe* (small life) for organisms – plant, animal and bacterial – that could be studied only with the aid of a microscope, but this was later reserved for bacteria and *micro-organism* introduced for microscopic life in the general sense.

As microscopes improved, the structure of the cell was progressively exposed. In 1831 Robert Brown gave the name *nucleus* to the small globule that occupied about 10 per cent of the volume of a cell. In 1868 Friedrich Miescher joined the University of Tübingen to work for Felix Hoppe-Seyler in the first laboratory in the world to be devoted to biochemistry, or physiological chemistry as it was then known. At that time proteins appeared to be the key to the chemistry of life – indeed the term is derived from the Greek for 'of first importance'. Miescher set himself the task of identifying the proteins in cells, and on examining cells from the thymus gland he found something unexpected: when treated using a weakly alkaline solution the nucleus would swell up, burst and release a substance that was not a protein, so he named this *nuclein*. He submitted his results to the laboratory's *Journal of Medical Chemistry* in 1869 but it was not published until 1871, after Hoppe-Seyler had isolated a similar substance from yeast. When Miescher later found nuclein to be a large molecule that incorporated acidic groups, it was renamed *nucleic acid*. Because the two samples had different chemical properties, Miescher's was named thymus nucleic acid and Hoppe-Seyler's yeast nucleic acid. It was initially thought that one pertained to animal life and the other to plant life, but this later proved to be naïve.

Meanwhile, in the 1860s the Austrian monk and amateur botanist G.J. Mendel kept a statistical record of his efforts to cross-breed pea plants. He found that each *trait* of the pea (e.g. the colour of its seed) was controlled by two *factors*, one donated by each parent, and also that while one factor may be *dominant* and the other *recessive*, the recessive factor was not lost as it often reappeared in a later generation. This formalised the concept of *inheritance*. Mendel published his results in 1866 in an Austrian journal, but attracted little attention.

To overcome the fact that the cell was essentially transparent when examined in a microscope, it was decided to utilise dyes to stain parts of a cell selectively. In 1879 the German biologist Walther Flemming discovered that a nucleus contains distinct granules, which he named *chromatin*. In 1881 Edward Zacharias found that chromatin was (at least partly) composed of nucleic acid. In 1884 Oskar Hertwig suggested that nucleic acid "is responsible not only for fertilisation but also for the transmission of hereditary characteristics". In 1888, after chromatin had been seen to form thread-like strands as it was shared in the process of cell division, Wilhelm von Waldeyer named the strands *chromosomes*. When Albrecht Kossel, another member of Hoppe-Seyler's laboratory, analysed salmon sperm, which are essentially

---

* Fungi are now regarded as being neither plant nor animal, but as a unique class of organism.

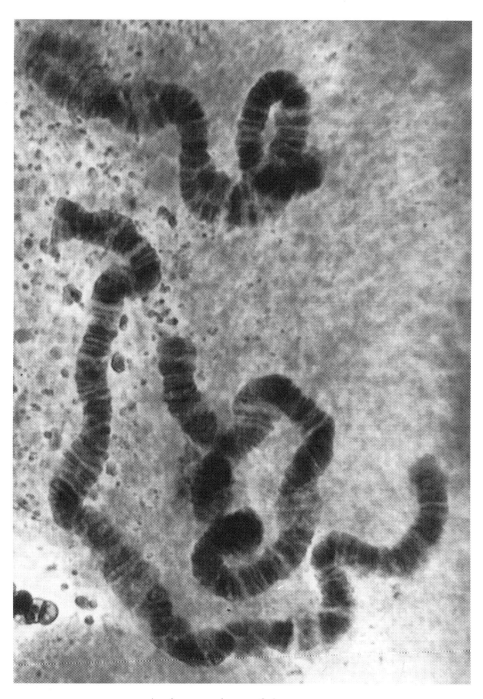

A microscope image of chromosomes.

just bags of chromosomes, he found that they contained twice as much nucleic acid as protein, and, furthermore, the protein in this case was a strikingly simple one. This did nothing to suggest that the hereditary information in the sperm cells was carried by the protein – the protein was suspected of being the 'stiffener' that held the chromosomes together. Later studies established that each species of plant or animal has a definite number of chromosomes, and that as a cell divides, the number of chromosomes doubles, the nuclear membrane dissolves, the chromosomes separate into two groups and a membrane forms around each group as the outer membrane is reconfigured to make two cells. Further investigation revealed that an egg or a sperm has only half the number of chromosomes for a cell of that organism, and the full complement is established when a sperm fertilises an egg. In 1896 the American biologist Edmund Wilson wrote in his book *The Cell in Development and Inheritance* that "we reach the remarkable conclusion that inheritance may, perhaps, be effected by the physical transmission of a particular compound from parent to offspring". The relationship between Mendel's factors and chromosomes was noted in 1904 by the American ctyologist W.S. Sutton, but it was evident that the number of possible traits greatly exceeded the number of chromosomes (of which there are 23 pairs for a human), which implied that each chromosome was a group of factors. In 1909 the Danish biologist W.L. Johannsen renamed Mendel's factors *genes* (from the root of 'genesis'). Unfortunately, although it appeared that nucleic acid was the key player in the chemistry of life, the focus of research then switched back to proteins.

BIOCHEMISTRY

Amino acids and proteins

An *acid* is a substance that readily yields a hydrogen ion, which can combine with a hydroxyl group (i.e. an oxygen and a hydrogen) as a water molecule. One of the simplest members of the *carboxylic acid group* is acetic acid, the main ingredient of vinegar. It is the carboxylic acid group that provides the acidity of organic acids. If ammonia loses one of its three hydrogen atoms by joining with a carbon chain, the result is an *amine group*. The *amino-carboxylic acids* – more commonly known as *amino acids* – have a structure in which a carbon atom links together:

- a hydrogen atom,
- an amine group,
- a carboxylic acid group, and
- a residual structure (R).

In the simplest amino acid, glycine, discovered in the 1820s, the residual structure is simply a hydrogen atom.

   In the first decade of the twentieth century the German chemist Emil Fischer suggested that amino acids linked up to form *polypeptide* chains, but it was not until the 1930s that it was established that proteins are polypeptides, as opposed to more complex structures. As in the case of polymers, the polypeptide chains are formed by condensation. That is, the amine group of one amino acid yields a hydrogen and the

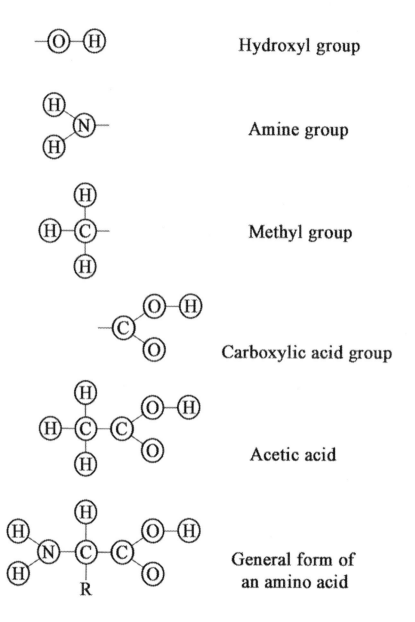

Hydroxyl group

Amine group

Methyl group

Carboxylic acid group

Acetic acid

General form of
an amino acid

The molecular configurations of the hydroxyl, amine, methyl and carboxylic acid groups, acetic acid, and the general form of an amino acid involving a residual (R).

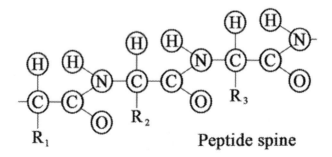

**Two amino acids form a peptide bond
and release water**

Two amino acids form a simple peptide bond and release a molecule of water. A
representation of the spine of a polypeptide featuring the sequence of residuals.

carboxylic acid group of the other yields a hydroxyl group, releasing water as a
byproduct of the creation of the *peptide bond*. There remains an amine group at one
end of the polypeptide and a carboxylic acid group at the other, ready for further
extension of the chain. The residuals express a sequence of 'side chains' which 'spell
out words'. In the 1930s biochemists set out to use X-ray diffraction to identify the
order in which the amino acids linked up to form the various proteins.

Attention returns to nucleic acids
Between 1883 and 1903 Kossel used hydrolysis to isolate from nucleic acids five
nitrogen-containing compounds, known as *bases*, which he named: adenine ('A'),
guanine ('G'), cytosine ('C'), thymine ('T') and uracil ('U'). While there are many

chemical bases with a variety of structures, those found in nucleic acids all have a ring structure: thymine, cytosine and uracil have a single ring and are members of the *pyrimidine* family, and guanine and adenine have the double ring of the *purine* family.

In 1911 P.A.T. Levene, a former student of Kossel, now at the Rockefeller Institute for Medical Research in New York, found that nucleic acids also contain five-carbon sugar molecules. In yeast nucleic acid this was *ribose*. In the case of thymus nucleic acid the sugar was almost identical except for the absence of one oxygen atom, which resulted in it being named *deoxyribose*. The nucleic acids were therefore renamed ribose nucleic acid and deoxyribose nucleic acid – terms that were promptly reduced to *ribonucleic acid* (RNA) and *deoxyribonucleic acid* (DNA). It was further found that although the nucleic acids each incorporated four bases, they differed in one of the pyrimidines, with DNA having thymine and RNA having uracil – which is identical to thymine except that the methyl group is replaced by a single hydrogen atom.

Meanwhile, the German chemist Robert Feulgen had established that DNA and RNA were both found in animals and plants – contradicting the early idea that one was associated with animal life and the other with plant life. It was known that the nucleic acid in chromosomes was DNA, but whereas there were only four bases in a

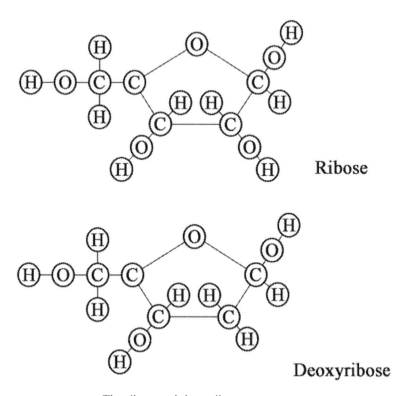

The ribose and deoxyribose sugars.

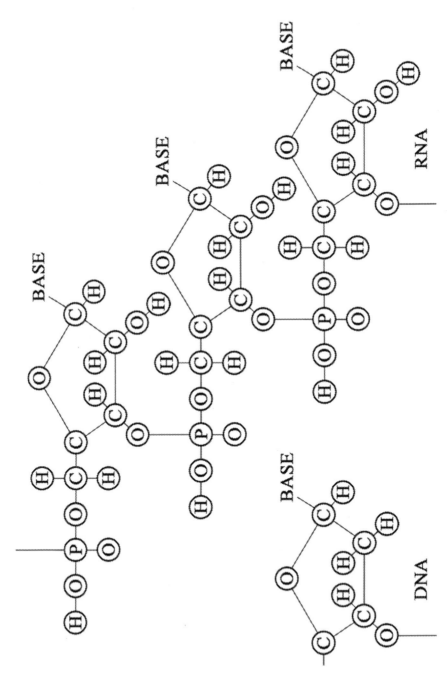

The phosphate–sugar backbone of ribonucleic acid, together with a fragment of the corresponding structure for deoxyribonucleic acid.

nucleic acid there were about 20 amino acids from which to make protein. It was therefore concluded – the earlier analysis by Kossel notwithstanding – that genetic information must be carried by proteins and the DNA was just a framework for the storage of proteins. It was reasoned that a cellular nucleus held the master copy of each protein, and that enzymes (themselves proteins) made copies when a cell required more protein. However, when Levene established that the nucleic acids could be assembled from *nucleotides* comprising:

- a base,
- either the ribose or the deoxyribose sugar, and
- a phosphate group comprising a phosphorus atom and four oxygen atoms

it was realised that nucleic acids are comparable in size to proteins.

A remarkable fact about hydrogen

The equations of state devised to describe the behaviour of liquids and gases were based on the assumption that the molecules were point-like and underwent perfect elastic collisions. In 1881 the Dutch physicist J.D. van der Waals modified the equations to allow for the fact that the molecules were large structures that could exert forces on each other when they came into close proximity. He did not know, however, what caused the force by which the molecules interacted, and it was not until the 1930s that the German physicist, Fritz London, having devised a quantum mechanical treatment of the hydrogen molecule, provided the explanation. The electron cloud around one molecule is attracted to the positively charged nucleus of another atom, and *vice versa*. This attraction is only significant when atoms are quite close, and is stronger when more electrons are present. The attraction is overcome by the repulsion between atomic nuclei only when atoms are close enough for their electron shells to merge. This is the basis for the differences between solid, liquid and gaseous states of non-ionic compounds. The molecules in a solid are held together, and are essentially static. If heated, they vibrate. If heated sufficiently, they are able to slip out of position and as they drift past each other the force slows their progress. If heated sufficiently for encounters to be rare, there is little opportunity for the force to apply, with the result that the molecules undergo elastic collisions – as assumed by the so-called perfect gas law. By this logic, the van der Waals force is stronger for larger molecules, so substances with greater molecular weights *should* have higher melting and boiling points. Water is an unusual liquid. Made of one oxygen and two hydrogen atoms it has a molecular weight of 18 daltons yet is liquid at room temperature while many heavier molecules are gases – for example, hydrogen sulphide weighs 34, carbon dioxide 44, and nitrogen dioxide 46. The van der Waals force would imply that water could *not* be liquid in conditions at the surface of the Earth, but it is, and remains so until heated to 100°C. The answer to this quandary is a force that is intermediate in strength between the covalent bonds that hold atoms in molecules and the van der Waals force between molecules.

Water is a V-shaped molecule with its two hydrogens separated by an angle of 104.5 degrees. The distribution of electric charge in this molecule is most strongly concentrated in the region between the hydrogen nuclei and the oxygen. There is

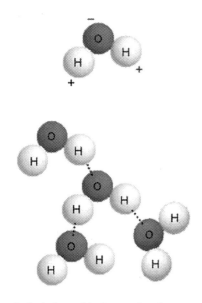

A depiction of hydrogen bonds between water molecules.

little charge on the opposite side of a hydrogen atom, where the single proton of its nucleus is barely screened – in effect, the hydrogen appears as a positively charged zone poking out through the overall electron cloud of the molecule. On the other side of the molecule from the two hydrogens is an overall negative charge, because the nucleus of the oxygen is shielded. Consequently, the molecule is *polarised*. As water molecules jostle each other, there is a force of attraction between them, with the hydrogens each attracted to the non-hydrogen side of another molecule. The result is that water remains liquid to higher temperatures than is permitted by van der Waals alone. Hydrogen is unique in having a single electron, and is the only atom that can act as the positive partner in this kind of bond between molecules – hence the term *hydrogen bonding*. There are several atoms that can form the negative partner in the bond – oxygen and nitrogen being particularly effective (i.e. hydrogen can form a bridge between two oxygens, two nitrogens, or between an oxygen and a nitrogen) – but carbon is *not* on this list. This might seem rather arcane, but life depends on it.

The structure of DNA

The *re*-discovery that it was DNA, not protein, that carried hereditary information came in 1944 when Oswald Avery at the Rockefeller Institute made a study of pneumococci, the agent of pneumonia, but further progress was impeded by the difficulty in isolating DNA for study. In the early 1950s, Francis Crick and James Watson in Cambridge set out to investigate the structure of DNA using X-ray diffraction data provided by King's College in London. In 1952 Crick speculated that the molecule was held together by the bases. John Griffith made a calculation and determined that (1) the shapes of the 'A' and 'T' molecules *could* fit together and link by a pair of hydrogen bonds; (2) the 'G' and 'C' molecules could link up by a trio of hydrogen bonds; (3) the two pairs were of the *same shape*; and (4) the bases *could not fit together in any other way*. Several months later Erwin Chargaff of Columbia University visited Cambridge and pointed out to Crick that samples of DNA always contained equal amounts of 'C' and 'T', and equal amounts of 'A' and 'G'. All of this clearly pointed the way, and by early 1953 Watson and Crick had shown that the phosphate–sugar groups serve as the spines of two intertwined helical molecules that are linked by spars formed from pairs of bases. Because in DNA the 'A' links up only to 'T', and the 'G' only to 'C', the order of the bases in one helix

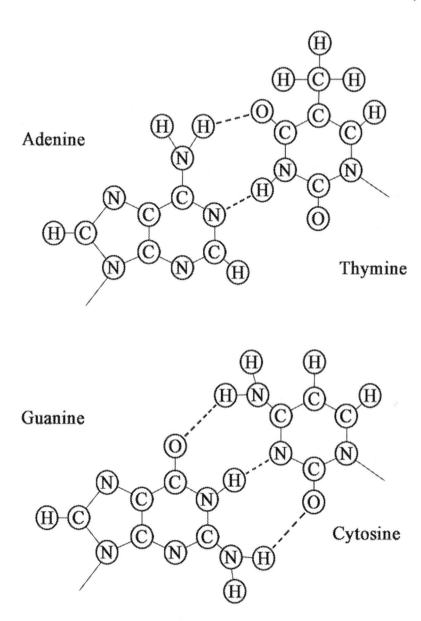

The hydrogen bonds (dashed) that link nucleic acid base pairs together.

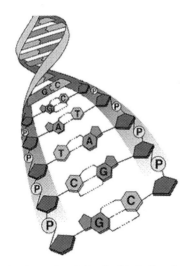

A schematic of the double helix of DNA with base pairing.

defines the order in the other, which is an ideal property for a molecule involved in self-replication. The 'genetic code' is a string of *codons*, each of which is represented by a triplet of bases and specifies one of the amino acids in the polypeptide chain of a protein.

RNA's crucial role

Whereas the amount of DNA in a cell remains constant and is the same in all cells of an organism, the RNA can vary in time and from one cell to another depending on how 'active' the cell is. Phosphate is not only used to link together the sugars of nucleic acids. Adenosine is adenine linked to ribose, and adenosine triphosphate is adenosine linked to a chain of three phosphate groups using bonds that store more energy than that between a carbon and a hydrogen atom, and are the primary means of storing biochemical energy.

In the 1940s Fritz Lipmann in America realised that not just adenine, but *any* base could be part of a phosphate-rich molecule, and the sugar could just as readily be deoxyribose as ribose. It was then noted that the components from which a cell produces a molecule such as RNA are not 'naked' bases but nucleotides with extra phosphates attached – that is, they carry their own supply of energy to overcome the barrier that would otherwise prevent them from attaching to the end of the chain. In the process, the phosphates are liberated and recycled – they are acting catalytically. A similar process provides the energy to join amino acids into a polypeptide chain to create a protein. In 1956 the electron microscope enabled G.E. Palade to discover that the sites of enzyme manufacture in the protoplasm were tiny particles rich in RNA, and so these were named *ribosomes*. A bacterial cell has typically 15,000 ribosomes, and a mammalian cell has 10 times that number.

In 1960 the French biochemists François Jacob and J.L. Monod realised that when a protein needs to be manufactured in a cell the appropriate part of the DNA uncoils and the string of three-letter codons for the gene is copied into a strand of RNA, and this RNA (which differs primarily in having 'U' where the DNA had 'T') passes from the nucleus out into the protoplasm. With this *messenger-RNA* at a ribosome, the protein is built up from the protoplasm, which is full of dissolved structures that are, in effect, triplets of bases attached to an amino acid. Because the triplet is unique to a given amino acid, as the triplets line up with those on the messenger-RNA their amino acids line up and link together. These structures have become known as *transfer-RNA* since each has a very short section of RNA – the triplet – on one end. The codon at that point in the messenger-RNA engages (by hydrogen bonding) the appropriate piece of transfer-RNA with the 'anticodon', in the same manner as in the double helix. Once the 'next' transfer-RNA settles into

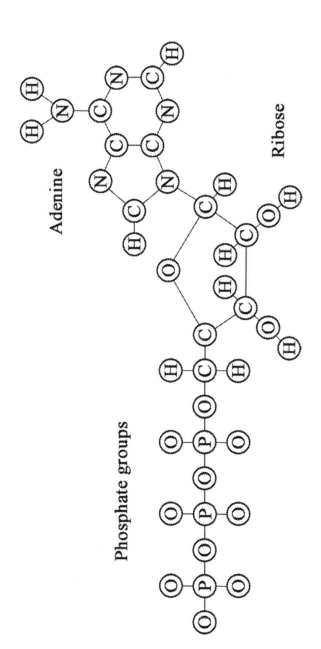

Adenosine triphosphate.

place, and its amino acid is in proximity to the end of the developing protein, an enzyme establishes the join, the ribosome releases the 'earlier' section of transfer-RNA, and this sets off to capture another amino acid from the protoplasm. Thus, as the ribosome advances along the messenger-RNA it builds the chain of amino acids for a specific protein. When a ribosome reaches the end of the messenger-RNA, it releases the complete protein, which another enzyme folds into a shape appropriate for its function in the cell. Several ribosomes can work simultaneously on a strand of messenger-RNA to manufacture many copies of the protein. When the messenger-RNA is no longer required, it is broken down and its components are reused. The ribosome is 'dumb' in that it reads the messenger-RNA irrespective of its sequence. A single population of ribosomes is therefore able to manufacture any (and all) proteins. It is a *very elegant* system.

The homogeneity of life

By the end of the twentieth century it had been shown that humans share 98.4 per cent of their genetic material with chimpanzees and gorillas. The fact that such a small percentage of genetic difference can yield such distinct creatures suggested that there were 'control genes' that regulated the behaviour of other genes. By 2001 the Human Genome Project had totally mapped the DNA of every human chromosome, with lengthy sequences identifying their genes. While it was not known what each gene *did*, it was clear that humans had about 30,000 genes. While sufficient to specify of the order of 250,000 proteins, this was considerably fewer genes than expected, being only twice that of a 'fruit fly' and only 4,000 more than a 'thale cress' weed. But the most profound discovery was that, in terms of DNA and the way in which cells manufacture proteins using messenger-RNA, there is *absolutely no difference* between the many forms of terrestrial life. Underlying the stupendous biodiversity there is a common biochemistry. How did this originate?

THE ORIGIN OF LIFE

In the 1950s S.L. Miller and H.C. Urey had performed pioneering experiments that synthesised simple amino acids from a reducing mixture of methane, ammonia and water vapour. The obvious next step was to demonstrate the linking of amino acids by peptide bonds. As earlier experiments had synthesised hydrogen cyanide, Juan Oro, in 1961, added this to the raw materials in order to 'kick start' the reaction, and synthesised a richer mixture of amino acids, a few short peptides, and some of the purines, particularly adenine. In 1962 he added formaldehyde and produced ribose and deoxyribose – the sugars essential to nucleic acids. When Ruth Mariner, Carl Sagan and the Celyonese biochemist Cyril Ponnamperuma added adenine to a ribose solution and irradiated it with ultraviolet, the result was adenosine. Adding phosphate yielded both adenosine triphosphate and the adenine nucleotide. In 1965 Ponnamperuma produced a dinucleotide of two linked nucleotides. In 1966 G.W. Hodson and B.L. Baker made the porphyrin ring that is the basis of chlorophyll. It seemed reasonable, therefore, to infer that DNA would be the inevitable result of such synthesis.

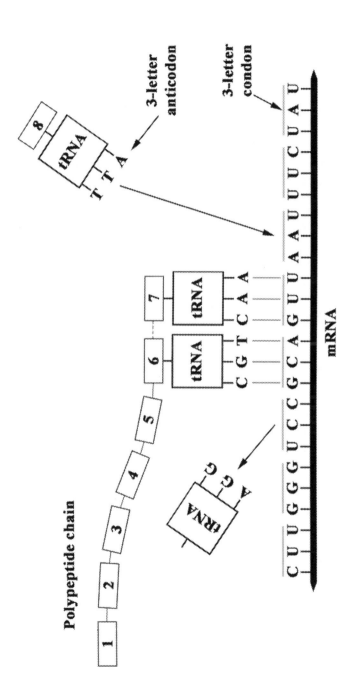

How a ribosome assembles a polypeptide chain of a protein.

The earliest evidence of life
The Earth accreted from the solar nebula some 4.5 billion years ago. Within about 100 million years its surface had cooled sufficiently to support a hydrosphere, but the bombardment of planetesimals continued to 3.8 billion years ago. Four billion years ago, the Sun radiated only 70 per cent of the energy that it does currently. As the Earth cooled, why didn't the water freeze and turn its surface into an icy crust? James Kasting, a climatologist at Pennsylvania State University, has proposed that the atmosphere would have been so pumped up with carbon dioxide emitted by the intense volcanism of that time that the 'greenhouse effect' kept the surface warm enough for water to remain liquid. Furthermore, whereas the ocean is an excellent store house for heat, at night continents lose most of the heat they have soaked up during the day. The early Earth was dominated by volcanic islands, and as there were no continents the ocean would have retained most of the heat arriving from the Sun. If there was a hydrosphere throughout most of this early era, when did life develop? The oldest known rocks are about 4 billion years old, but they have been altered by subsequent processing, and hence can reveal nothing of the existence of life at that time. Nevertheless, the existence of life cannot be ruled out. The 3.8-billion-year-old gneisses in Greenland that were laid down in water provide a hint of life, but metamorphism has deformed and heated the rocks – which is why they are gneisses. The identification of 'biomarkers' in a variety of ancient rocks was repeatedly reported and then dismissed: amino acids proved to be contamination, and putative 'microfossils' were globules of fluid that had become trapped inside crystals as minerals grew. However, as organisms preferentially assimilate carbon-12 over carbon-13, sediment made of cellular debris is consequently enriched with carbon-12 and carbonate precipitated from water that is depleted of carbon-12 is enriched with carbon-13. When carbonate is heated under pressure, it turns into graphite. Analysis of the isotopic compositions of carbon-12 and carbon-13 seeks evidence of oceanic life *indirectly*, by its chemical signature. Unfortunately, the signal in the Greenland gneisses is weak, and disputed. By 3.5 billion years ago, however, the evidence for life is much more convincing: well-preserved rocks at Barberton in South Africa and in the Pilbara region of western Australia contain evidence of complex microbial ecosystems – including fossilised stromatolites, microfossils of cells and chemical signatures. Stromatolites are stacks of calcareous sediment in irregular pancakes that were made by colonies of 'blue-green algae', a single-celled *cyanobacteria* that developed the ability to utilise chlorophyll for photosynthesis. These formed thin mats on the floors of shallow seas and layers of rock built up by accretion as the mats trapped particulates in suspension and minerals precipitated from the water. While the biological origin of the fossil stromatolites in the Pilbara has been disputed, there is no doubt about stromatolites at Baffin Island in Canada, which are 1.8 billion years old and can be matched point for point with the 'living fossils' that survive in the highly saline parts of Shark Bay on the western coast of Australia.

Archea
In the 1970s, when Viking was being built to test for microbes on Mars, biologists discovered a new domain of life on the Earth! One strand to the discovery of what

became the *archea* (the old ones) was of *thermophillic* microbes in hot springs. It had been believed that water in the hottest geothermal springs must be sterile, but in the early 1970s a microbe was found living in water at 85°C in a spring in the Yellowstone National Park. In 1977 a submersible that was investigating the rift in the Galapagos Ridge found a hydrothermal vent that was emitting a super-heated plume of water. A similar vent found on the East Pacific Rise in 1979 was so rich

Living stromatolites in Shark Bay, Western Australia.

in dissolved minerals that precipitates had constructed a 'chimney' above it. These 'black smokers' support ecosystems involving hundreds of species of life. While many species were related to clams, mussels, shrimps and tube worms, others were new to science. In 1982 it was found that these isolated food chains were based on single-celled organisms that derived their energy from nutrients emerging from the vent. A 3.26-billion-year-old sediment in western Australia hosts exceptionally well-preserved mineral textures that are indistinguishable from contemporary black smokers. Perhaps life developed in this environment?

Autotrophs versus heterotrophs
Biological cells are either *autotrophs* or *heterotrophs*, or both, in that they either manufacture organic compounds themselves from raw ingredients such as carbon dioxide (autotrophs, meaning self-feeders) or they assimilate organics from their environment and process these into what they require (heterotrophs). Which came first was a matter of debate. The heterotrophic origin hypothesis was proposed in the 1920s by the Russian chemist A.I. Oparin, and seemed reasonable. After all, why shouldn't the earliest life have exploited the organic compounds present in its environment? Modern opinion has, however, swung in favour of an autotrophic origin, which was reinforced by the discovery of yet another unusual microbial ecosystem. In the mid-1980s, microbes were found living in the interstices between grains of rock at a depth of 1 kilometre beneath the surface of Colorado's Piceance Basin. There were some heterotrophs present that consumed the remains of plant detritus bound up in sedimentary rock. They resembled micro-organisms living on the surface, but had adapted to a hot anoxic environment. Faced with such a limited supply of energy, they tended to remain immobile in their individual niches between rock grains. Most of the microbes, however, were autotrophs that exploited the heat and hydrocarbons that were toxic to 'conventional' life. Because they 'lived off rock', these were named *lithotrophs*.

Autotrophs are well represented among the archea. Those that thrive in a hydrothermal environment lacking in both oxygen and light are collectively known as either *anaerobic autotrophs* or *chemolithotrophs*. Many require only water enriched with volcanic gases and nutrients. The fact that *none* of the archean autotrophs makes use of sunlight implies that photosynthesis developed later. The development of the ability to use sunlight was a great advance, since it is more *efficient*. No longer restricted to sites of hydrothermal activity, cyanobacteria were free to colonise the planet. They photosynthesised carbon dioxide and water into carbohydrates and liberated oxygen, and on death 'decayed' by reacting with oxygen and reverting to carbon dioxide and water. In principle, because this was a reversible reaction, they should have had no effect on the environment, but over a 200-million-year interval some 2.2 billion years ago the fraction of oxygen in the atmosphere rose rapidly to about 15 per cent. A sudden change in the ratios of the carbon isotopes shows that a large amount of organic carbon became locked up in rock at that time. Once in anoxic sediment, the organics would have been unable to decay, and this tipped the equilibrium of the reaction in favour of an increasing oxygen concentration, which facilitated the development 1.5 billion years ago of the *eukarya* that were able to exploit the even more efficient process of respiration to oxidise organic compounds to produce the energy for enzymatic processes. The eukarya are characterised by having their genetic material isolated in a membrane-bound nucleus. A eukaryotic cell also contains other packages, collectively known as *organelles*, which include the *plastids* (centres of photosynthesis) of algae and plants, and the *mitochondria* (centres of respiration). A crucial discovery was that plastids and mitochondria were once *free-living bacteria*. It is likely that the larger cells developed as symbiotic relationships facilitating an *intracellular* division of labour. The idea was proposed early in the twentieth century, but was ridiculed. In the late 1960s it was revived as 'endosymbiotic evolution' by the American biologist Lynn Margulis and only recently proved by genetic studies. Once multi-cellular eukarya developed, the increased scope for mutation drove an 'explosion' in the diversity of eukarya about 600 million years ago.

The 'tree of life'
Only a few species of microbe have had their DNA completely sequenced, but one small information-rich molecule in ribosomal RNA that is found in *all* organisms has been studied. After surveying in excess of 100 species that were representative of all known major forms of life, a 'universal tree of life' was drawn up to shed new light on the origin of life. This technique was pioneered by Carl Woese of the University of Illinois, starting in 1966, but it only really made an impact in the late 1990s when it became efficient. The results provided a comprehensive chart of the evolutionary relationships showing that terrestrial life is predominantly microbial. In fact, of the three great domains of life – bacteria, archea and eukarya – most are microbial. Animals and plants that formed the focus of earlier classifications form the tip of just one branch of this tree. Anyone who is interested in the origin of life but studies only *macroscopic* eukarya is missing the point! Most of the eukarya are *microscopic*. Archea and bacteria (all of which are microscopic) were once lumped together as the *prokarya* – a domain characterised by the absence of a nucleus and organelles, and by asexual reproduction – but in fact archea have more in common with eukarya than bacteria, *suggesting that the ancestral eukaryotic cell was an*

archeon. It was the archea that organised DNA into a nucleus and engulfed bacteria in a large membrane to form a symbiotic structure. Nevertheless, archea differ from *both* bacteria and eukarya. Fully 56 per cent of the 1,738 genes in the first archeon to be completely sequenced are different to any in the other domains – reinforcing their classification as a distinct domain.

The 'root' of the tree of life was recently identified by a genetic study. The idea was that if some gene in the population of organisms ancestral to *all* terrestrial life duplicated itself, and the duplicates evolved, then those two genetic lineages could be compared to find the point of divergence. Three such genes were identified, and the root traced. The organisms nearest the root (as presently known) use hydrogen to reduce carbon dioxide to supply their cell material; that is, they are autotrophic. All of the species near the root of the tree are *hyperthermophiles* that grow *best* at high temperatures. The earliest organism was probably a thermophillic autotroph. It is therefore likely that life exploited geothermal energy – either on the surface or on the sea floor. Hot springs provide thermal and chemical energy, and are rich in nutrients leached from the rocks through which the water passes. The food chain of a black smoker ecosystem is based on microbes that gain their energy from the oxidation of the hydrogen sulphide that emerges from the vent. Although the water can be as hot as 400°C, it does not boil because the pressure in the rift, typically at a depth of 3 kilometres below sea level, is 300 bars and this prevents bubbles from forming. The black smokers are therefore much hotter than a hot spring on land. The dissolved chemicals precipitate on emerging into the cold water of the ocean. In excess of 100 of these very hot vents have been found, and there are many more that emit water at 200°C. Such vents would have been much more common on the early Earth. The hyperthermophiles thrive at temperatures in the range 100–200°C. In the hottest vents, they live not in the main flow but in the surrounding water. It was once thought that organic molecules would break down in superheated water, but the early atmosphere was predominantly carbon dioxide (estimates range from 100 to 1,000 times as much as now) and the surface pressure would have been 10 times greater than now, and this pressure would have inhibited the dissociation of such molecules.

All of the above notwithstanding, anaerobic autotrophs may *not* have been the first form of life to develop; genesis may have begun with 'proto-organisms' that were not cellular in form.

The 'RNA world' hypothesis

In 1897 Ronald Ross, a British physician in India, followed up a suggestion made three years previously by Patrick Manson, a colleague in Hong Kong, that the mosquito may have something to do with malaria, and Ross traced the disease to a non-bacterial micro-organism that the mosquito carried as a parasite. In 1899 the American bacteriologist Walter Reed identified yellow fever as being caused by another non-bacterial parasite carried by the mosquito. While the parasite causing malaria was a protozoan, the one causing yellow fever was not – it was something much smaller than even a bacterium and therefore something new. The Dutch bacteriologist M.W. Beijernick introduced the term *virus* (poison in Latin) and suggested that it was a chemical agent, but it proved to be larger than most organic molecules. In 1935 W.M. Stanley in America adapted a protein-separation technique to concentrate a virus, and was surprised that in its pure form it was crystalline (nothing living had previously been observed in this state), yet when redissolved in water it was just as virulent as previously, which suggested that it was *more dead than alive*. A virus seemed to be something 'non-living' that was capable of reproducing, but only when it invaded a living cell. In 1936, by which time some 40 diseases had been shown to be viral, F.C. Bawden and N.W. Pirie in Britain established that one virus, although mostly protein, also contained a small percentage of RNA, and this was later shown to be generally true. In fact, viruses were found to be composed of the same 'stuff' as genes. Most viruses are isolated chromosomes of up to several dozen genes and when they invade a cell *their* genes commandeer the metabolic apparatus normally controlled by the nucleus, often killing the cell and in some cases even the entire organism. In 1967 T.O. Diener found that the smallest viruses are 'naked' strands of RNA, and proposed the name *viroid*. A viroid as short as 400 nucleotides is still capable of replication. It has recently been suggested that the primordial ocean contained a 1 per cent solution of organics, which, as synthesis experiments have shown, would have included ribosomes. Life may therefore have begun with 'proto-organisms' that were strands of RNA that replicated, or not, depending on the organic compounds they met in the chemical soup, of which they were a part. The next step was the development of a membrane for a cell, and the simplest cell wall is made of peptidoglycan, which is a single large polymer of amino acids and sugar. It is therefore possible that an 'RNA world' preceded the emergence of the anaerobic autotrophs, and whether or not this constituted 'life' is essentially only a matter of definition.

5

A multiplicity of missions

THE SOVIETS

In May 1971 the Soviet Union launched the Mars 2 and Mars 3 missions, each of which was to drop a probe into the planet's atmosphere. As the landers were to be released several hours before the main spacecraft entered orbit, there was no way to revise either the time of landing or the target site. By sheer bad luck, one of the most extensive dust storms ever observed was raging when they attempted to land on 27 November and 2 December. After entry, the probe was to discard its conical heatshield, employ a parachute to make a 'semi-hard' landing, and then open four petals to expose its instruments. Nothing was received from the first probe, but the other touched down at 25 metres per second and its mothership relayed 14 seconds of contrast-free television comprising only a few scan lines. It was later found that the autonomous trajectory correction by Mars 2 six days prior to its arrival set the approach hyperbola too low, which caused its probe to penetrate the atmosphere at a very steep angle and hit the surface before its timer could release the parachute. In view of the stable configuration of the deployed lander, it is highly unlikely that Mars 3 was blown over. One suggestion is that the relay system on the mothership failed. Another is that electrical interference from the dust storm either disrupted the radio transmission or caused a discharge that zapped a critical system. The first mothership entered the planned orbit ranging between 1,280 and 25,000 kilometres but its radio link was so weak that little useful data was received. As a result of a propellant leak, Mars 3 limped into an orbit with its apoapsis 10 times higher than intended, severely limiting its opportunities for observing the planet, which was, in any case, obscured by dust, and by the time the storm abated the pre-programmed sequencer had completed its activities.

The Soviets tried again in 1973, but because this window was less favourable they had to split the landers from the orbiters and

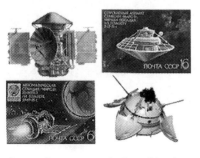

A set of stamps mark the orbiter/lander mission of Mars 3.

dispatch them separately. Mars 4 and 5 were to enter orbit, and Mars 6 and 7 were to release landers during flybys, with the landers' signals being forwarded by the orbiters. Unfortunately, each of the vehicles suffered a congenital defect in its transistors that proved fatal. Although the problem was discovered shortly prior to launch, there was no time to rectify it, and it was decided to proceed and hope that all would be well. As Mars 4 approach the planet on 10 February 1974 a propellant leak precluded entering orbit, but the spacecraft snapped a few pictures as it flew by at a range of 2,200 kilometres. Two days later Mars 5 entered the desired synchronous orbit. When Mars 7 arrived on 9 March it released its lander slightly early, causing it to miss the planet by 1,300 kilometres. On 12 March Mars 6, having taken a slower route, released its probe with Mars 5 standing by to relay the signal. Unfortunately, like Mars 3, it fell silent on reaching the surface. Shortly thereafter, the instrument compartment on Mars 5 lost pressure, pre-empting its mapping mission.

After the VeGa missions had delivered landers and upper-atmosphere balloons to Venus *en route* to a rendezvous with Halley's comet in 1986, the Soviet Union decided to send two such spacecraft on an ambitious mission to orbit Mars and investigate Phobos, the larger of the planet's two moons. These were successfully dispatched in July 1988 but, unfortunately, a commanding error caused the loss of Phobos 1 in late August. A sequence, some 20 or 30 pages long, had been uplinked to the spacecraft to be executed while it was out of communication with Earth, but the final digit was omitted and the spacecraft's computer interpreted this as an order to deactivate the attitude control thrusters. In the absence of positive attitude control, the vehicle started to rotate, lost solar power, and was never heard from again. Its partner entered Mars orbit on 30 January 1989, but was lost on 27 March during a manoeuvre in close proximity to its objective. In fact, these spacecraft had been doomed prior to launch, as their computers had components that were known to be

An artist's depiction of Phobos 2 manoeuvring near the larger of Mars's moonlets.

unreliable. One processor on Phobos 2 expired *en route* and a second began to malfunction intermittently soon after the craft entered orbit around Mars. Since the 'voting logic' was not reprogrammable, when the two faulty units voted 'no' and the surviving processor was ignored, attitude control was lost.

Meanwhile, planning had started in 1988 for another extremely ambitious idea. As originally envisaged, this would exploit the 1992 window to launch two large orbiters using the same kind of bus as the Phobos spacecraft, one of which would release a pair of small landers and a pair of penetrators, and the other would dispatch a rover and a French-supplied balloon sonde that would drift at low altitude during the day and settle onto the surface at night. However, budget problems in 1990 resulted in one spacecraft being rescheduled to August 1994 and the other to November 1996.

The Marsokhod rover had three axles mounted on a 1.5-metre-long 'backbone' that could be extended to vary the distance between the axles in order to deal with different terrains. Each of the six conically shaped titanium wheels was driven by its own electric motor. The articulated system was designed to enable the vehicle to clamber over a 1-metre-tall rock. The final design was to be 70 kilograms, have a payload of 15 kilograms, drive for 1 hour per sol at speeds of up to 1 kilometre per hour, and be energised by a radioisotope thermal generator. A prototype controlled by an umbilical was tested in 1991 on volcanic terrain in Kamchatka, the objective being to assess its basic mobility. In early 1992 a shortage of funds in the recently independent Russian Republic threatened the development of the first spacecraft, which was to deliver the landers and penetrators. If the 1994 window was missed, both would go in 1996, but since *that* window was less favourable the payload on the delayed mission would have to be scaled back. Later in 1992 a version of the rover with a radio-command link was tested at Death Valley in California, to assess how well it was able to operate autonomously. In early 1994 this was fitted with a robotic arm supplied by McDonnell Douglas, and tested for a week on the Kilauea volcano in Hawaii by remote control via a satellite link from the Ames Research Center in California. When it was confirmed that the first spacecraft would not be ready in time, its launch was slipped to 1996 and, since the development of the first mission was consuming the start-up funds for the second, it became financially impractical to send the rover and balloon in 1998, and it was decided to send the rover at the next window, in 2001, and the balloon in 2003. Although the company was unable to pay their salaries, the Lavochkin employees worked round the clock through the summer of 1996 to ensure that Mars '96 would be ready, since delaying it to 1998 would impose a further reduction of its payload.

Mars '96 was an international venture with participants from 20 countries and 30 American co-investigators. The 50-kilogram 'small station' soft landers were to be released a few days prior to the spacecraft's arrival at Mars in September 1997. At the end of its parachute descent, the lander would inflate two airbags to cushion the impact. On the way down, a French-supplied imaging system would document the site. Once on the surface, the 1-metre-diameter lander would deploy four petals – in the manner of its predecessors. The instruments included a Russian panoramic camera, a magnetometer supplied by France, meteorological instruments supplied by

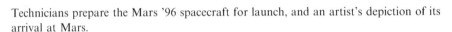

Technicians prepare the Mars '96 spacecraft for launch, and an artist's depiction of its arrival at Mars.

Finland, an alpha-particle, proton and X-ray spectrometer supplied by the Max Planck Institute in Germany to determine the chemical composition of the surface, and an experiment devised by Gilbert Levin to test for the presence of superoxides in the soil in order the reassess the results of the Viking biology package. Powered by small radioisotope thermal generators, the stations were to operate for 160 days. Their data was to be relayed to the orbiter over a French-supplied communications link. The approach trajectory required that both landers be aimed at sites just to the west of Olympus Mons. After entering orbit, the main spacecraft was to release the penetrators – one in the vicinity of the landers and the other on the opposite side of the planet. Each 65-kilogram penetrator was 2 metres in length. The aft part was to remain on the surface while the forward part, linked by an umbilical, penetrated to a depth of several metres with a seismometer and temperature sensors to monitor the subsurface heat flow for one Martian year.

After a scrubbed launch on 12 November, the Proton successfully lifted off on 16 November and the fourth stage entered the planned parking orbit at an altitude of 160 kilometres. Shortly thereafter, the vehicle was to make a 2,850-metre-per-second burn to produce a 100,000-kilometre apogee, at which point the spacecraft was to separate, fire its own engine, and head for Mars. Unfortunately, the stage failed to orient itself properly for the burn, and although the control system automatically shut down the engine, the brief burn caused the orbit to decay, with the result that the spacecraft fell into the atmosphere.* This débâcle prompted the cancellation of all future plans.

* If it had been successfully dispatched, the spacecraft would have become Mars 8.

THE AMERICANS

Mars Observer

In the mid-1980s NASA decided to reduce the cost of its missions within the inner Solar System by developing an Observer bus that would reuse electronics created for the Satcom 4000 series communications satellites and DMSP 5D2 and Tiros-N Earth-monitoring satellites. To further reduce costs, only one vehicle of each type would be built. The first mission, Mars Observer, was launched on 25 September 1992 to conduct a global survey of the surface and atmosphere over the course of a Martian year using a high-resolution imaging system; a laser altimeter to construct a topographic map; gamma-ray and thermal emission spectrometers to chart the composition of the surface; an infrared radiometer to monitor the annual cycles of the atmosphere; and an integrated magnetometer and electron reflectometer to seek magnetic fields. Unfortunately, contact with the spacecraft was lost on 22 August 1993, three days before it was due to orbit Mars. Its delicate transmitter had been deactivated as a precaution against the shock from firing the pyrotechnic valves that were to open the propellant lines in preparation for the orbital insertion burn. As contact was never regained, the investigation was unable to identify the fault conclusively, but it was suspected that a slow propellant leak, which would have been inconsequential on a satellite that used its main propulsion system only for a few days in manoeuvring to its operating station, had build up during the 11-month interplanetary cruise. In addition, in retrospect it was realised that because Mars is twice as far as Earth from the Sun, the thermal environment would have been very different to that envisaged when the propulsion system was designed. In view of this loss, the idea of reusing systems developed for Earth satellites was abandoned.

An artist's depiction of Mars Observer in its fully deployed state in orbit around Mars.

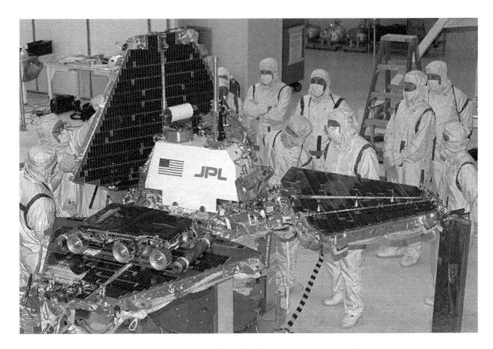

Technicians prepare the Mars Pathfinder lander with its Sojourner rover.

The rationale behind Mars Observer had been to equip a single spacecraft with instruments to make coordinated studies of a wide range of phenomena – a strategy that would pay off handsomely if events went to plan, but would result in the loss of all of the science if the spacecraft were to suffer a catastrophic failure. Adopting a new strategy of 'faster, cheaper, better', NASA decided on a long-term Mars Exploration Program that would exploit every launch window over a period of a decade to send a series of smaller probes, each equipped to pursue a specific investigation. However, prior to the loss of Mars Observer, another project was in development for the 1996 window, this being for the Discovery Program, which was to demonstrate low-cost (i.e. $150 million) planetary missions.

Mars Pathfinder

Mars Pathfinder was dispatched on 4 December 1996 and reached Mars on 4 July 1997. As with the Soviet landers, it was to enter the atmosphere from the approach trajectory. The primary engineering objective was to test an entry, descent and landing system utilising – in turn – a heatshield, parachute, rocket thrusters and airbags. A radar altimeter was to inflate the airbags at an altitude of 300 metres, the thrusters were to fire at 50 metres to slow the rate of descent, and the lander was to be released. While the still-firing thrusters drew the chute clear, the lander would fall, bounce, and roll to a stop. Shortly thereafter, the bags were to deflate with the material drawn beneath the open petals of the lander, out of the way. The requirements were that the site be low-lying for maximum parachute braking, and in

An orbital view of the Mars Pathfinder landing site. (see page 185 for larger context)

the equatorial zone for solar power. The system was designed to enable a small lander to be delivered to a site that was much rougher than a Viking could have tackled. A site was selected where the Ares Vallis outflow channel debouched onto Chryse Planitia. As this was rough terrain – about 850 kilometres southeast of the Viking 1 site, and not far from the site originally chosen for that lander – it was hoped to find a litter of rocks that had been swept in from a wide variety of sources, including the highlands.

The lander, which was named the Carl Sagan Memorial Station, was equipped with a stereoscopic camera and a suite of meteorology instruments, but its primary payload was Sojourner, an experimental 6-wheeled rover having an alpha-particle, proton and X-ray spectrometer similar to that developed for the ill-fated Mars '96 mission. Whereas the X-ray fluorescence spectrometers on the Vikings could only analyse soil scooped by the robotic arm, Sojourner was able to manoeuvre to place its instrument against individual rocks. Also, as the light-travel time meant that the rover could not be controlled from Earth in real-time, its computer was capable of operating in a semi-autonomous manner. The first rock it examined was 'Barnacle Bill', so-named because of the large knobs on its surface. The result was surprising: a high silica content suggestive of a 'dry' andesite. The second rock, named 'Yogi', was a low-silica basalt. If taken at face value, this meant that Mars was a more thermally evolved planet than had been thought, which raised the issue of when its internal 'heat engine' had ceased. A visual inspection of the textures of the rocks revealed that some of the boulders were conglomerates in which pebbles were bound in a fine matrix, suggesting they had formed when a suspension settled out after a flood. Cavities in the conglomerates showed where pebbles had eroded out; there were pebbles lying around that may have been shed in this manner, and their rounded form suggested that they had undergone considerable erosion *prior to* their being conglomerated. Furthermore, the fact that the conglomerates were delivered to this site intact implied that they had been formed either in a river bed or in the aftermath of an *earlier* flood. Other conglomerates, in particular the large angular fragments, were more likely to be ejecta from the impact that formed 'Big Crater', located several kilometres to the south. The Vikings had shown the soil to be rich in sulphur, and Sojourner confirmed this, strengthening the theory that global dust storms have homogenised the dust planetwide. Despite the suggestion of andesite, the analyses indicated that the fine material was weathered basalt. In addition to confirming the Viking finding of a highly magnetic mineral in the soil, the new lander's magnetic properties experiment established this to contain maghemite – a mineral that, on Earth, forms when an iron-rich aqueous solution is freeze-dried – and this supported the hypothesis that the planet was once warm and wet. With the benefit of hindsight, the Sojourner team realised that the rocks were coated with a rind of weathered material that was not chemically representative of the rock beneath. Although it was generally agreed that the site marked an ancient flood, this did not prove that the planet had once had a hydrological cycle, as an intense but brief eruption of water from the surface could have occurred even in a cold and dry climate. To prove that the planet had been warm and wet over a geologically significant interval, it would be necessary to locate outcrops of sedimentary rocks.

The Sojourner rover inspects one of the large rocks near the lander.

The Pathfinder lander exceeded its one-month baseline mission, and went on to send weather reports at 100 times better temporal resolution than had the Vikings. The hopes of monitoring the onset of the northern winter were frustrated when the lander fell silent on 27 September 1997, very likely because its battery froze in the night, at which time Sojourner, having greatly exceeded its one-week mission and studied eight rocks and two soil samples, was denied its communications relay.

Mars Global Surveyor
Following the loss of Mars Observer, it was decided to divide that mission's many instruments between a series of smaller spacecraft, the first of which, Mars Global Surveyor, was launched on 7 November 1996 and arrived on 12 September 1997. It was to use a similar Sun-synchronous mapping orbit, but attain this in a different manner. Whereas Mars Observer would have fired its engine a number of times to manoeuvre into the desired orbit over a period of three months, the new spacecraft was to use its engine to enter an initial 'capture' orbit with a high apoapsis, and on reaching this point fire its engine to lower the periapsis into the upper atmosphere in order to use 'aerobraking' to progressively reduce the apoapsis over a six-month period, at which time the periapsis would be lifted. The rationale was to reduce the mass of propellant required, so that the spacecraft could be launched on a smaller, and cheaper, rocket. The objective was to collect data for at least one local year using upgraded forms of the high-resolution camera, laser altimeter, thermal emission spectrometer, and integrated magnetometer and electron reflectometer that had been developed for Mars Observer.

Aerobraking must be undertaken cautiously, because the drag force depends on the density of the upper atmosphere, which can vary with time, location and solar activity. The altitude must balance dipping deeper into the denser atmosphere for rapid braking against the ability of the spacecraft to tolerate the resulting structural and thermal loads. Mars Global Surveyor's solar arrays were to cant at an angle of 30 degrees, with their rear surfaces facing into the wind to provide 'weathercock' aerodynamic stability. However, when the solar arrays were deployed early in the interplanetary cruise, one failed to latch into place. Each array had two panels, and it was suspected that the input shaft of the viscous damper that was to prevent the hinge overshooting had sheared, possibly because the inner panel was still moving when the outer panel locked into position. A Sun sensor indicated that the angle of the panel had changed during the cruise, suggesting that the yoke, which was a triangular epoxy–aluminium honeycomb assembly that connected the panel to its gimbal, was tearing away from the panel. As the aerobraking loads would exceed the torque of the spring that held the array in position, the unlatched panel was rotated by 180 degrees in the hope that the air pressure during the first aerobraking pass would force it out and latch it into position. Although, this would expose the solar cells to the free molecular flow heating, tests established that this would not impair their efficiency. Once the panel had locked into place, it would be rotated to undertake the subsequent aerobraking as planned. On 17 September the spacecraft fired its engine to enter a capture orbit ranging between 110 and 54,000 kilometres. On an aerobraking pass on 6 October deceleration loads were found to be 50 per cent greater than expected, with indications that the damaged solar array had been bent *beyond* its deployed position. Controllers raised the periapsis to keep the spacecraft above the atmosphere while the problem was studied. It was eventually decided that the damaged panel could take a dynamic pressure of only one-third of that planned.

An artist's depiction of Mars Global Surveyor with its solar panels configured for aerobraking.

The only option was to perform the aerobraking at a higher altitude in order to reduce the drag, which would make the process much more protracted. When the periapsis was lowered again on 7 November, the aerobraking proceeded without incident and the spacecraft finally achieved the required mapping orbit in March 1999. As a consolation prize, the extra time spent with a low periapsis had enabled the magnetometer to collect considerably more high-resolution data than it otherwise would have done, revealing the surface to be magnetised in a manner that raised intriguing issues regarding the planet's early history.

A double blow

Continuing its 'faster, cheaper, better' strategy, NASA built an orbiter and a lander to investigate Mars's climate. In spring, the south polar cap of carbon dioxide frost retreats, exposing a 'layered terrain' characterised by alternating bands believed to represent different mixes of dust and water-ice. It was hoped that remote sensing by Mars Climate Orbiter and *in-situ* sampling by Mars Polar Lander would enable the structure of this terrain to serve in a manner analogous to 'tree rings', and show

An artist's depiction of Mars Global Surveyor in its fully deployed state.

whether the planet has undergone cyclic variations in climate over approximately the last 100,000 years.

Although flown after Mars Pathfinder, the development of Mars Polar Lander started before the airbag system was demonstrated, and therefore for entry, descent and landing it was to employ the proven technique of a heatshield, parachute, and rocket thrusters. It was launched on 3 January 1999, and made a direct entry on 3 December. A downward-facing camera was to document the site, in order to assist in interpreting the surface sampling – in particular, to indicate the type of 'layer' it was on. Because there was no facility for in-flight telemetry this imagery was to be recorded and transmitted from the surface. Since the lander was solar powered, the landing site had to be as far north as possible, and a 'tongue' of the layered terrain that projected 15 degrees (800 kilometres) from the pole at 195 degrees longitude was selected from which the cap would have retreated several weeks previously.

In addition to a stereoscopic camera to survey the site, Mars Polar Lander was to excavate a trench using a 2-metre-long arm which had its own camera to inspect the wall for fine layering. The arm was to supply up to eight subsurface samples to an instrument that would measure the 'volatiles', in particular water and carbon dioxide that were physically and chemically bound into the soil. After completing its sampling activities, Mars Polar Lander was to serve as a meteorological station for three months, monitoring the winds, the concentration of water vapour, and the altitudes of water-ice clouds and dust hazes.

The landing site for Mars Polar Lander, and artist's depictions of the vehicle firing its engines for the terminal phase of its descent and later using its robotic arm to sample the surface.

In addition to an infrared radiometer – left over from Mars Observer – for profiling the atmosphere, Mars Climate Orbiter had a new colour-imaging system capable of taking horizon-to-horizon views at medium resolution (40 metres per pixel) in order to monitor the weather globally on a daily and seasonal basis. The investigation was to focus on water, in particular clouds, frost on the surface, and the daily and seasonal variations of water vapour. The spacecraft was launched on 11 December 1998 on a trajectory to arrive on 23 September 1999. As the capture orbit would not be as eccentric as in the case of Mars Global Surveyor, it would be possible to complete the aerobraking to attain Sun-synchronous orbit rapidly. This spacecraft had a different configuration to its predecessor, with just one solar array comprising three panels that projected to one side like a wing. After serving as the primary communications relay for Mars Polar Lander, it was to monitor the polar caps and atmosphere for one Martian year.

Unfortunately, Mars Climate Orbiter was lost while firing its engine behind the planet to enter the capture orbit. Owing to a simple unit-conversion error, the craft had dived too deeply into the atmosphere and burned up. A data file sent from the contractor to the JPL engineers who were to control the spacecraft's trajectory was supposed to have been expressed in metric units, but was instead in Imperial (so-called 'English') units. This introduced a discrepancy factor of 4.45, which resulted in errors that caused the spacecraft to fly too close to the planet. The first sign of trouble was when the vehicle slipped behind the planet's limb 49 seconds early. When the trajectory was recalculated taking into account the units error, it was realised that the vehicle had penetrated the atmosphere to an altitude of 57 kilometres, which was too deep for survival.

A second blow was delivered when Mars Polar Lander arrived several months later, was released by its cruise stage, and was never heard from. The investigation

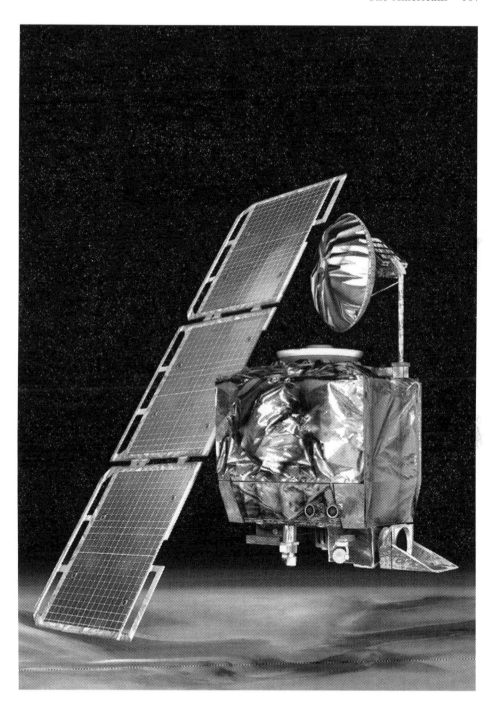

An artist's depiction of Mars Climate Orbiter.

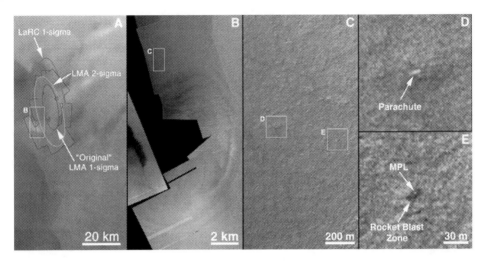

After a lengthy search, Mars Global Surveyor identified the site where Mars Polar Lander crashed.

was hindered by the fact that the lander was not equipped to issue telemetry during entry, descent and landing. The failure mode was discovered several months later. The parachute was deployed at an altitude of 7 kilometres. Ten seconds later, the heatshield was jettisoned and the lander's three legs deployed. The radar acquired the surface at an altitude of 1,500 metres. With 40 seconds to go, the backshell was jettisoned. The lander ignited its engines, cancelled its horizontal drift and started to regulate its sink rate. At a height of 40 metres, having achieved the required rate of descent, the computer started to monitor a signal that was to indicate that one of the legs had made contact with the surface, but a design flaw that should have been caught in testing had resulted in this being asserted in the act of deploying the legs, prompting the computer to shut off the engine immediately. As the fall would have lasted just several seconds, the lander would likely have come to rest in an upright position, but the shock would almost certainly have damaged the waveguide of the communications system that was affixed to a sidewall, rendering it useless.

Mars Odyssey

Following the tragic loss of Mars Climate Orbiter, orders were issued to preclude a repeat of the confusion over units of measurement, and procedures for the capture orbit and aerobraking were made more conservative. It proved possible to dispatch the next orbital mission, but the planned lander was deleted while surface activities were reassessed. As the next window occurred in 2001, the new orbiter was named Mars Odyssey in homage to Arthur C. Clarke's novel *2001: A Space Odyssey*. It was launched on 7 April 2001, arrived on 24 October, completed its aerobraking in January 2002 and began its primary mission, which was to last at least one Martian year. It addressed the 'follow the water' theme at the heart of the Mars Exploration Program using two instruments to study the surface. The gamma-ray spectrometer

An artist's depiction of Mars Odyssey.

(the final instrument from Mars Observer) was to measure the abundances of 20 elements in the topsoil at a spatial resolution of 300 kilometres – sufficient to give a sense of the general character of the planet. It had been augmented with neutron detectors with which to sense hydrogen, indicating the presence of either hydrated minerals or water-ice in the uppermost metre or so of ground. The vehicle also had a new instrument, in the form of a thermal emission spectrometer integrated with a medium-resolution optical imager to correlate the mineralogical survey with the associated landforms. It was capable of detecting carbonates, silicates, hydroxides, sulphates, oxides, and hydrothermal silica in the topsoil at abundances of about 10 per cent, and was to make a global mineralogical map at a resolution of 100 metres per pixel to enable structures associated with the action of water to be identified, and so give insight into the past climate. If there were active hydrothermal systems that broached the surface, the thermal imager would be able to detect them against the chilly surface in the night. As the first orbiter capable of detecting near-surface ice outwith the polar regions, Mars Odyssey was expected to shed light on where the ancient water went.

EUROPE JOINS IN

In early 1999 the European Space Agency gave Matra Marconi Space the contract to build the spacecraft for its first mission to Mars. This "fast, flexible and cheap"

Technicians prepare Mars Express, and an artist's depiction of the spacecraft in its fully deployed state.

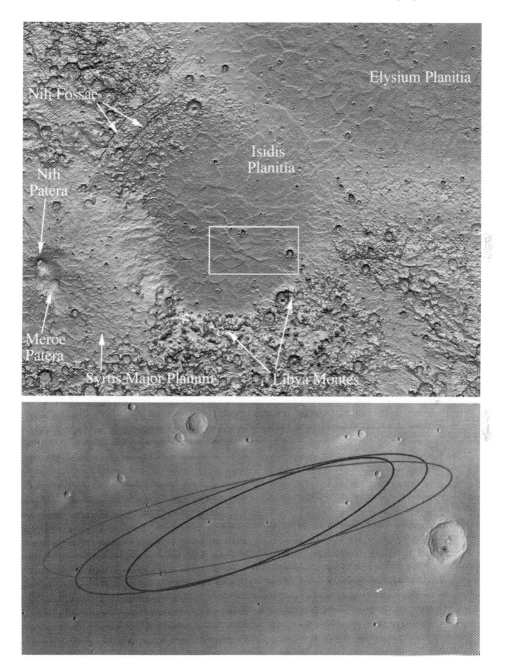

A regional view of the Isidis basin, and the ellipse of the landing site selected for Beagle 2.

spacecraft would share technology with the Rosetta cometary mission, and employ instruments developed for the ill-fated Mars '96 – namely a high-resolution colour imager, an infrared spectrometer for mineralogical mapping, two spectrometers to measure the atmospheric composition on a local scale to study circulation patterns, and an instrument to investigate how the solar wind interacts with the atmosphere. In addition, it had a new instrument: a long-wavelength radar able to penetrate the ground to a depth of several kilometres in search of water, either liquid or ice. The primary British contribution was a lightweight lander devised by Colin Pillinger of the Open University, called Beagle 2 in homage to *HMS Beagle*, the ship on which Charles Darwin made his epic voyage in the 1830s. Mars Express was launched by a Russian Soyuz–Fregat on 2 June 2003.

Beagle 2 renewed the search for signs of life, either extinct or still extant. It was to inspect rocks for minerals that would indicate the presence of liquid water in the past, seek carbonaceous structures left by organisms living in that water, and measure the ratio of the carbon isotopes – a 'biomarker' test that has been used to investigate when life developed on Earth. As regards the possibility of extant life, this lander was *not* to make a Viking-style test designed to promote metabolic activity; instead it was to analyse the atmosphere for out-of-equilibrium gases that could be the result of metabolism by, for example, methanogenic archea. Unlike the Vikings, which had analysed samples taken from topsoil irradiated by solar ultraviolet, solar wind and cosmic rays, this lander was to investigate more benign environments by extracting cores from rocks and using a 'mole' to retrieve a sample from a depth of 1.5 metres, where the soil would not only be shielded but might also be moist. To test the hypothesis that superoxides were being produced in the atmosphere by the ultraviolet flux and accumulating in the topsoil, it was to measure the ambient ultraviolet insolation, the rate of oxidation by the atmosphere, and the oxidation state of the iron in the soil. As in the case of Mars Pathfinder, the landing site had to be low-lying for parachute braking and near the equator for solar power. Isidis Planitia was selected as it was a large sediment-filled basin with what appeared to be a smooth surface.

An artist's depiction of Beagle 2 on the surface of Mars.

Mars Express released Beagle 2 on 19 December to pursue entry, descent and landing autonomously on 25 December, but the signal to indicate that it had landed was never received. Since there was no facility for in-flight telemetry, its fate is a mystery, but it is possible that a recent dust storm caused the upper atmosphere to thin to such a degree that the parachute was ineffective.

6

The water dilemma

TWO HEMISPHERES

The impression of Mars gained from the initial flybys was of an ancient cratered plain that had undergone little erosion, but between them these probes had imaged barely 10 per cent of the surface. When Mariner 9 mapped the planet from orbit, it revealed a much more diverse landscape. To a first approximation, there are two morphologically distinctive hemispheres. The transitional 'line of dichotomy' has its most northerly point at about 330 degrees longitude at 50 degrees latitude. The smooth plains to the north may be either effusive volcanism or some other form of sedimentation on a vast scale, and are comparatively lightly cratered. The rims that protrude in places indicate that this infill, whatever its nature, smothered a cratered terrain. Significantly, the northern plains are several kilometres below the 'datum', defined as the elevation at which the atmospheric pressure is 6.2 millibars; this being the 'triple point' for water. The boundary is characterised by an irregular but shallow scarp that is scalloped where major impacts created basins that were later flooded from the north. It is not yet known whether this hemispheric dichotomy is due to endogenic or exogenic processes. In addition, Mariner 9 revealed that Mars has huge volcanoes and channels, suggesting that the planet once had a dense atmosphere, and was warm and wet.

VOLCANISM

The Tharsis province

Measured in terms of how it rises from the surrounding terrain, the Tharsis bulge has an asymmetric profile 4,000 kilometres north to south and 3,000 kilometres east to west. Its steep northwestern flank rises from the northern plains and forms a series of young lava flows, but the shallow eastern flank, which is a ridged plain, transitions into the ancient cratered terrain.

Alba Patera, the first volcano to develop in the province, lies on the northern flank of the bulge. Although its base is 1,500 kilometres across, it is an extremely shallow edifice whose summit rises only a kilometre or so. Concentric fractures at the

An elevation map of Mars as determined by the laser altimeter on Mars Global Surveyor.

periphery indicate that the structure is in an advanced state of collapse. Early in its history, the volcano was probably an explosive vent that blanketed its environs with pyroclastics. In Mars's low gravity, an ash cloud would have been blasted to an altitude of 100 kilometres, and the plume would have collapsed and surged over the surface. In fact, this ash is still visible in a ring between 250 and 450 kilometres from the caldera. The eruption style then became effusive, and sheets of lava were fed by channels and tubes that can be traced up to 1,000 kilometres west from vents near the irregular 100-kilometre-wide caldera complex.

The broad summit of Tharsis is 8 to 10 kilometres above the datum. Syria Planum is intensely incised by the intersecting canyons of Noctis Labyrinthus. There are long fractures radiating down the flanks of the bulge, and others forming concentric arcs on the lower slopes. In many cases, slabs of crust have dropped between pairs of faults to form grabens several kilometres across. Deeper faults on the eastern flank created the canyon system of Valles Marineris, which extends one-quarter of the way around the equator. The bulge also hosts a line of shield volcanoes – Arsia, Pavonis and Ascraeus Mons – spaced about 700 kilometres apart, whose lava flows have partially engulfed many of the much smaller volcanic cones and grabens on the blanks of the bulge.

To the northwest of the bulge is Olympus Mons, the largest volcano on Mars. Its lower slope is truncated by a scarp that matches the elevation of Syria Planum. The structure is a stack of lava flows, some of which, particularly in the northeast and the southwest, are draped over the scarp; in other places the scarp is an almost sheer cliff, but it follows an irregular line and short sections are oriented radially to the edifice. The slope above the scarp is only a few degrees, but about one-third of the way up its profile becomes a series of 10-degree ramps onto broad terraces that were

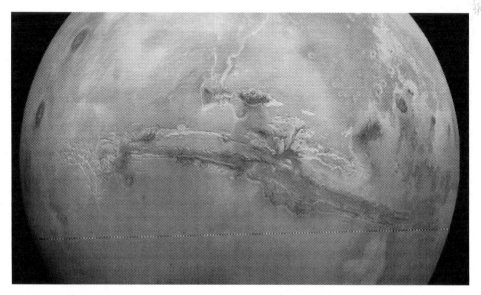

A mosaic of orbital imagery featuring the canyons of Valles Marineris.

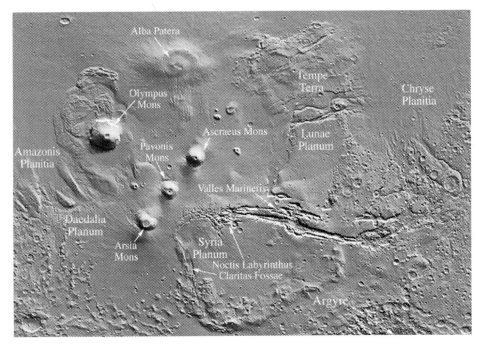

An elevation map of the Tharsis province.

probably formed by slippage on shallow thrust-faults as the edifice inflated and deflated in response to magma rising into and draining from the upper magma reservoir – a process known as 'breathing'. The summit is 17 kilometres above the crest of the scarp. The 80-kilometre-wide caldera complex is marked by a cliff that drops 2 kilometres to the multifaceted floor, which records at least six phases of activity. There are large lobate features of an uncertain origin several hundred kilometres out from the scarp. The ridges and grooves imply that these deposits are several kilometres thick. It has been suggested that they are the material that slumped off Olympus in a 'base surge' and exposed the scarp. As the three large shields sit on the 10-kilometre-tall bulge, their summits are 27 kilometres above the datum. The summit of Olympus is at this elevation, but because it stands on terrain that is just 2 kilometres above the datum it is a larger edifice, 550 kilometres in diameter. The fact that all of the large shields peaked at the same elevation suggests that growth from their summit vents ceased when the pressure in the lithosphere was no longer able to force new magma up the feed pipes. After the summits of the shields on the bulge became dormant, the residual pressure opened parasitic vents on their lower flanks. Magma rising in a volcano's feed pipe is driven by the hydrostatic pressure induced by the difference in density between the magma and the rock it must pass through. Where these densities were comparable to terrestrial values, the magma feeding these volcanoes would have been drawn from a depth of 250 kilometres. As the lithospheric 'plate' that forms the floor of the Pacific Ocean migrated across a mantle 'hot spot', magma from a depth of 60 kilometres rose to

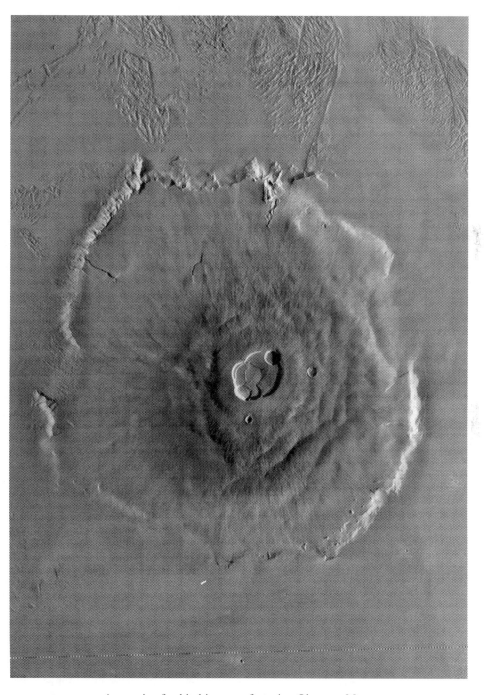

A mosaic of orbital imagery featuring Olympus Mons.

create the succession of volcanoes of the Hawaiian chain. Even if all this magma had formed a single edifice, it would not have rivalled Olympus. However, it must be borne in mind that Olympus has been building for a very long time. The great bulk of the Martian shields derives from the fact that the lithosphere is immobile.

The Elysium province
There are three modest shields in the Elysium area, which rises 5 kilometres above the low-lying northern plain. The first was Hecates Tholus, some 6 kilometres tall. Activity shifted 850 kilometres south to make Albor Tholus, a 30-kilometre-wide dome. Later, the 10-kilometre-tall shield of Elysium Mons formed between them, and its lava not only encircled the older edifices but also ran 1,300 kilometres east onto the cratered terrain.

Highland volcanism
There are also volcanoes in the southern highlands around the periphery of Hellas, which, at 2,000 kilometres in diameter, is the largest of the well-preserved impact basins. There is an arc of mountains to the north and west, the southeastern rim is degraded, the southwestern rim is masked by volcanism, and the northeastern rim is etched by two channels that drained down the shallow slope to deposit material on the floor which, despite such infill, lies 7 kilometres below the datum. Being 3.3 kilometres above the datum, the surrounding ejecta is the highest terrain in the southern hemisphere. The reduction in pressure on the lower lithosphere by the removal of crustal material during such an impact would have stimulated localised decompressional melting, and deep faults would have allowed magmatic intrusions to reach the surface, where extremely low-profile structures with complex calderas were created. Situated 1,500 kilometres northeast of Hellas's rim, Tyrrhena Patera is highly degraded. A series of explosive eruptions imply that magma rose through water-saturated megaregolith and the water flashed to steam. The flanks are deeply eroded with radial channels, suggesting that pyroclastic flows left thick blankets of ash, which later 'welded' into ignimbrites. The complex caldera indicates episodic volcanism, and it has been proposed that once the vent had exhausted its supply of volatiles it switched to effusive lava, and that it was this, rather than water, that eroded the deep channels. One of the most prominent channels emerges from the caldera, runs west for more than 200 kilometres, and merges with the surrounding volcanic plain. If the ash was etched by water, however, this is more likely to have been by sapping induced by the local heat flow through the crust melting ice rather than by the runoff of prodigious precipitation. Hadriaca Patera, which is nearby, also underwent explosive eruptions that deposited thick blankets of ash. Indeed its subdued caldera produced an apron 300 kilometres across that was so thoroughly etched by radiating channels that it is now strikingly ridged. Given the location of Hadriaca on the regional slope, its lava cut channels 400 kilometres long, draining to the southwest. The nearest terrestrial analogue for the explosive phase of a Martian patera is Yellowstone in Wyoming, a 'resurgent caldera' that repeatedly blanketed a large part of the North American continent with ash. With flanking slopes of just a fraction of a degree, Amphitrites and Peneus Paterae southwest of the rim of Hellas

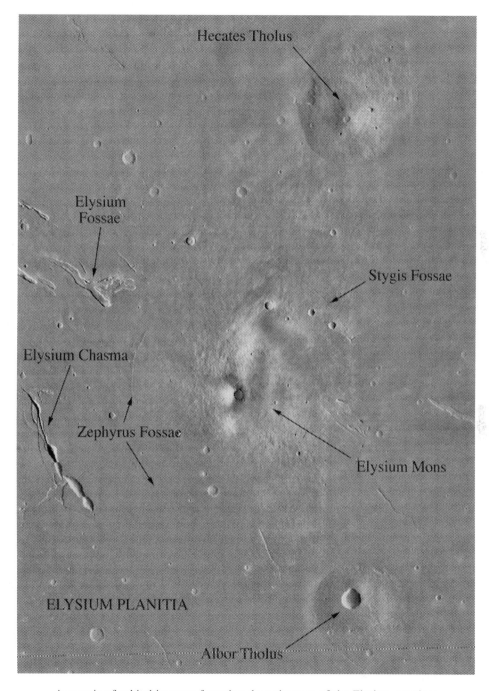

A mosaic of orbital imagery featuring the volcanoes of the Elysium province.

The heavily eroded Tyrrhena shield volcano.

are so subdued that their calderas resemble impacts on open ground, but their effusive lavas embayed the ancient cratered terrain.

Apollinaris Patera is located in the transitionary terrain just beyond the line of dichotomy. It may have initially issued ash, but is believed to be the earliest of the lava shields. The western flank is surprisingly steep. The final phase of activity left the 70-kilometre-diameter caldera full of lava, with an extrusion spilling down the southern flank. Nili and Meroe Paterae on Syrtis Major Planum represent another type of highland volcanism. They resemble ultra-low-profile shields on a volcanic plateau, and exceed 1,000 kilometres in width. Arcuate grabens in the central area trace out a shallow depression 280 kilometres across. This volcanism appears to be related to the nearby Isidis basin in the same way that Tyrrhena Patera is related to Hellas, because the vents lie on arcuate fractures concentric to the basin. There is a similarly subdued caldera-like feature named Tempe Patera on cratered terrain to the northeast of Tharsis. There are no such low-profile central-vent structures on the low-lying northern plains; the conditions that led to their formation evidently only arose in the ancient cratered terrain.

MORPHOLOGICAL EVIDENCE OF SURFACE WATER

Valley networks

There are several types of valley networks in the southern highlands. The runoff channels are so like terrestrial drainage systems that the case for their having been etched by slowly running surface water is compelling. They are typically less than 1 kilometre wide, seldom exceed 100 kilometres in length, start small and increase in size downstream, have dendritic inflow channels, and often end abruptly just as if the water had disappeared underground. On Earth, the erosion that forms karst in permeable rock often creates such 'blind alleys'. There are also structures that look as if they formed by collapse when permafrost melted and water erupted from the ground. Such valleys are more or less straight and have steep walls. The tributaries are very short, join high up on the walls, and would have discharged by waterfalls. It is thought that such a valley would have been progressively extended by sapping at its head, with the tributaries forming in the same manner, as sapping eroded the cliff-like wall. This interpretation is supported by the fact that the tributaries emerge from amphitheatre-like cavities that show no sign of having served as runoff collectors. As an integral feature of the southern highlands, the valley networks clearly derive from an early era, possibly a time when the atmosphere was much thicker, pumped up by greenhouse gases emitted by the volcanoes, and the climate was warmer and wetter. Gullies in the walls of valleys and craters suggest that water seeped out and ran downslope – although when this occurred is a matter of debate.

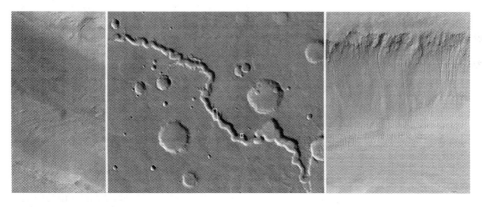

A general view of Nirgal Vallis and two images by Mars Global Surveyor showing sand dunes on its floor and gullies in its wall.

Outflow channels

The most startling of Mariner 9's revelations were what appeared to be enormous outflow channels. Apart from two on the eastern rim of the Hellas basin and some on the flank of the Elysium rise, these channels drain across the line of dichotomy onto the low-lying northern plains. Most of these channels debouch onto Chryse Planitia. This vast drainage system comprises Ares, Tiu and Simud (all of which emerge from crustal collapse structures, typically 100 kilometres across, known as

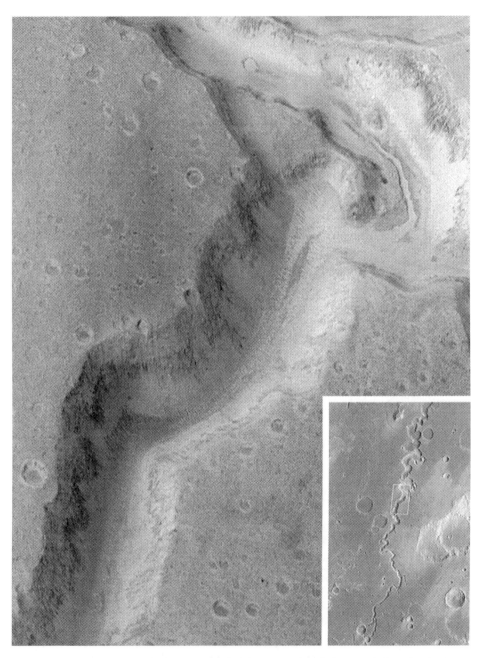

This view of Nanedi Vallis by Mars Global Surveyor revealed not only layering in the wall but also an intriguing inner channel alongside a bench at the corner which looked as if it had recently been cut by flowing water.

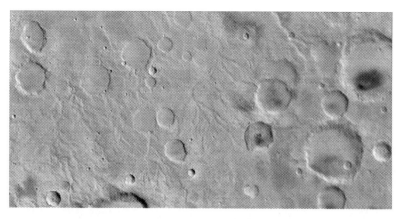

A dendritic valley system on the intercrater plain of the southern highlands.

'chaotic' terrain), Shalbatana (which derives from a smaller chaotic zone north of Eos and crosses Xanthe Terra), Maja (drawn from Juventae Chasma on the eastern side of Lunae Planum), and Kasei (which emerges from Echus Chasma to the west of Lunae Planum and runs north before swinging east into Chryse). Although such channels emerge from their sources fully formed, some have short tributaries that emerge from their own chaotic terrains. It has been suggested that these sources formed on ring fractures that are concentric to the impact basin underlying Chryse Planitia, the permafrost was melted by magma that exploited these deep faults, and once this water found an outlet and drained the 'pore space' the surface collapsed. On the other hand, it has been argued that the flooding from these chaotic zones is on such a vast scale because it was driven by the 'break out' of an aquifer that was 'pumped' by gravity-driven artesian flow in the Tharsis bulge. These floods were extremely erosive, sculpting isolated obstacles on the plain into teardrop shapes. They swept away sections of the ejecta blankets of craters and incised their flanks to leave them standing on 'pedestals', and, if obstructed by a ridge, accumulated until the water level reached the crest and etched a deep channel as the dam drained. On debouching onto the low-lying plains, they spread out and deposited the suspended material, forming gravel bars through which later flows eroded 'braided' channels. A close terrestrial equivalent, at least in terms of how the landscape was carved, is the Scablands of eastern Washington State, which was formed when an ice dam in northern Idaho broke at the end of the last Ice Age and released a large glacial lake which, in a matter of days, drained a volume of water equivalent to that of Lakes Erie and Ontario.

Ancient sediments
Mars Global Surveyor's high-resolution camera found evidence of sedimentary rock layers inside craters, between craters and in deep chasms. There were several types of outcrop. Layered units comprised thin rock beds, often just several metres thick, that were neatly stacked in distinct groups. Massive units were bulkier, with no clearly defined bedding. In some cases, massive units were found together with a layered

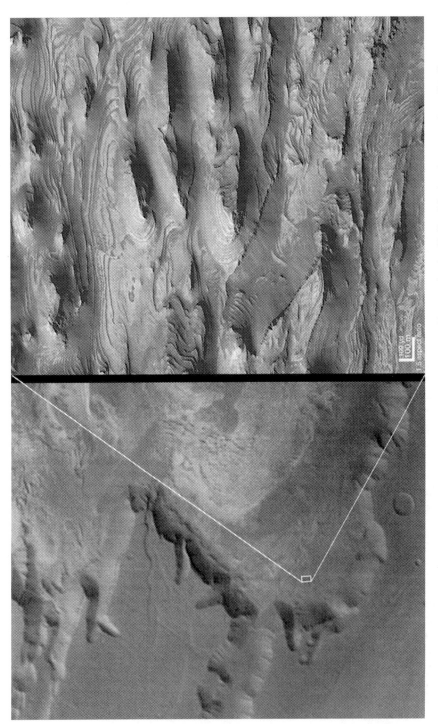

Mars Global Surveyor revealed layering on the floor of West Candor Chasma in the Valles Marineris suggestive of sedimentary deposition in ancient lakes.

unit, although in every case the massive unit was sitting on top. The caps of eroded massive or layered units were shallow mesas with surfaces ranging from smooth to pitted, ridged and grooved. All such outcrops appeared to be fine-grained materials in horizontal layers. In some places hundreds of individual beds were exposed, and in a few cases alternating light and dark deposits were stacked several kilometres thick. The surrounding terrains indicated that the deposits were laid down very early in Martian history, between 4.3 and 3.5 billion years ago. While it is possible that these deposits were produced during intense dust storms when the atmosphere was denser, it seems more likely that they were formed in standing water because, although the exposures occur at a variety of locations, they are most common in isolated sites such as canyons and craters in which water could be expected to have pooled. The absence of drainage channels into these locations has been interpreted as indicating that subterranean water was driven to the surface by artesian pressure, as the local 'water table' rose. The confined nature of the sites implies that the water stood as lakes for a significant time, and eventually evaporated.

An ancient river delta

In November 2003 Michael Malin and Kenneth Edgett reported finding an ancient sedimentary distributary fan (a generic term that includes river deltas and alluvial fans) in a 64-kilometre-wide crater northeast of Holden in the southern highlands. This fan-shaped apron of debris, some 13 kilometres in length and 11 kilometres in width, is downslope of a large network of channels that apparently drained into the crater. The fan was estimated to contain about one-quarter of the material removed during the cutting of the channels. This discovery was hailed as the 'smoking gun', showing that in the distant past it was possible for liquid water to flow across the surface in a persistent way. The lithified sediments now stand above their eroded surroundings as curved ridges that trace the meanders created as the flowing water changed its course over time. The general shape, the pattern of the channels, and the shallow local slopes suggested that the feature was actually a delta – a deposit left where a river enters a body of standing water. If so, as Malin said, "it would be the strongest indicator yet that Mars once had lakes".

The Athabascan 'ice sea'

An outflow of water from the Cerberus Fossae on the southeastern flank of the Elysium rise carved Athabasca Vallis as it drained south. In addition to etching the terrain with tell-tale teardrop-shaped landforms, this flood created 'megaripples' – which are, in fact, the best preserved case of ripples formed in a catastrophic flood anywhere on the planet. An 800-kilometre-wide area nearby that was patterned with irregular plates raised the intriguing possibility that this water is still present in the form of a cluster of icy raft-like structures that, as viewed from orbit, bear a striking resemblance to terrestrial ice floes. In the proposed scenario, the water started to freeze as it ran downslope, and the floating pack-ice broke up into irregular rafts. Being located near the equator, the ice ought to have melted and the water sublimed, but the fact that it did not implies that the ice was soon covered, most likely by volcanic ash, in a deposit that was thick enough to insulate the ice and yet not so

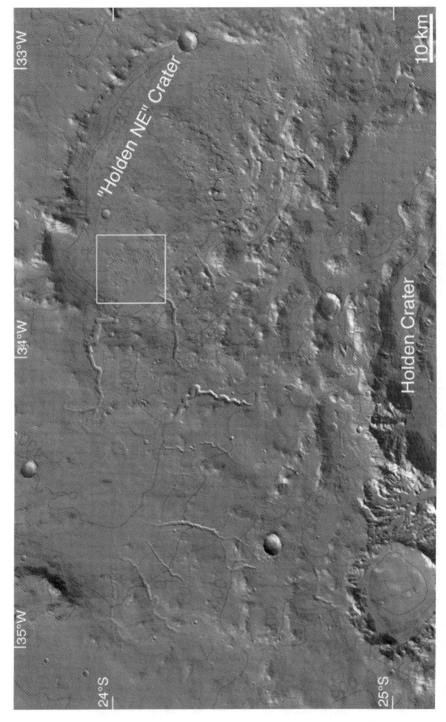

The area north of the crater Holden in the southern highlands shows a multitude of features suggestive of erosion by surface water.

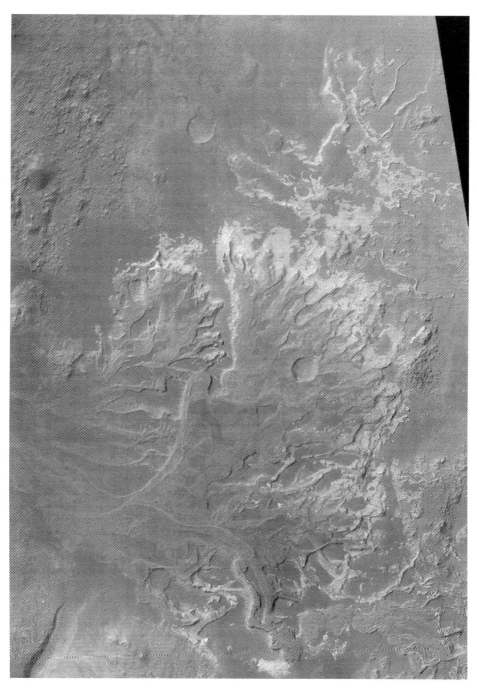

The eroded relic of a sedimentary distributary fan north of the crater Holden.

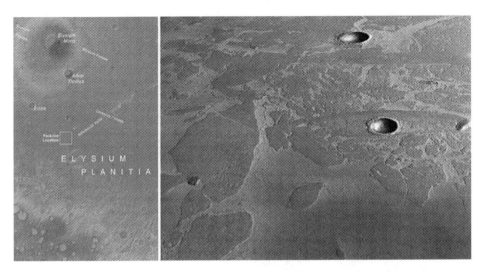

This patterned ground has been interpreted as rafts of ice made when water from Cerberus Fossae flowed down Athabasca Vallis, pooled in low-lying ground, froze and was mantled by a shallow blanket of ash.

thick as to mask the surface relief. The level surfaces suggested that ice is still present under this veneer, and the thickness of the ice was estimated by studying the craters on the rafts. In announcing the finding at the conference in the Netherlands in February 2005 to discuss the first year's results from Mars Express, John Murray of the Department of Earth Sciences at the Open University in England estimated that the ice covered an area of 750,000 square kilometres to a depth of 45 metres, and was therefore comparable to the North Sea on Earth. Furthermore, the crater counts indicated that the structure was recent, possibly less than 5 million years old.

"I think it's fairly plausible," said Michael Carr of the US Geological Survey at Menlo Park in California. "We know where the water came from; you can trace the valleys carved by water down to this area."

Oceanus Borealis
In 1986, when studying Viking orbital imagery, Timothy Parker of the University of Southern California began to notice subtle features on the northern plains that bore a strong resemblance to the margins of a lake that had flooded parts of Utah, Nevada and Idaho 100,000 years ago. In 1989, having integrated the evidence, he suggested that a series of oceans with differing quantities of water had formed and retreated north of the line of dichotomy. It was suggested that these formed when magma intrusions melted permafrost. The Mars Orbiter Laser Altimeter (MOLA) on Mars Global Surveyor, which could measure elevations to within 1 metre, found that the elevation of the northern plains varies by no more than about 100 metres over distances of hundreds of kilometres. The profiles along the line of dichotomy were reminiscent of the continental slopes that sweep down onto the abyssal sedimentary plains of Earth's oceans. The greater of two shorelines proposed by

A pattern of megaripples in Athabasca Vallis.

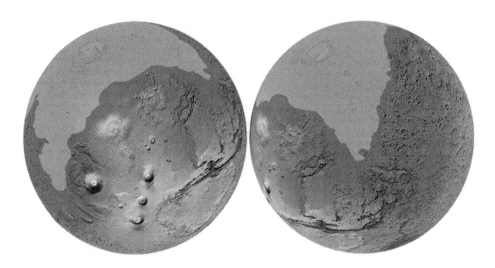

A depiction of the Oceanus Borealis according to the 'inner contact' which some researchers have interpreted as an ancient shoreline. (Courtesy of Peter Neivert of Brown University.)

Parker, the so-called Arabian shoreline, at an elevation 2 kilometres below the datum, was found to follow an undulating elevation. However, the entire circumference of the 'shallower' Deuteronilus shoreline, 3.5 kilometres below the datum, traced out a single contour remarkably well, deviating by no more than 280 metres from its mean, and where the 'fit' was poor the reason was obvious: on Elysium Planitia, for example, it had been overrun by lava flows. To James Head of Brown University, Rhode Island, the data provided four types of quantitative evidence favouring the existence of an ancient ocean within this line: (1) it is almost a level surface; (2) the topography is more subdued below this level than above it, which is consistent with smoothing by sedimentation; (3) a series of interior terraces are highly suggestive of a receding shoreline; and (4) the volume contained within the shoreline is consistent with estimates of the amount of water available on the planet.* However, other researchers have pointed out that the terraces have the characteristic profile of 'wrinkle ridges', and may be tectonic structures produced by regional compressive stresses. When Mars Global Surveyor returned high-resolution imagery of parts of the putative shoreline, there was little on the surface to support the contention. On the other hand, it might be that, on a world dominated by eolian erosion, an ancient shoreline will be so eroded that it is difficult to identify close up and is best inferred from circumstantial evidence. For example, Head has pointed out that over a distance of 2,200 kilometres six major outflow channels debouched onto Chryse within 180 metres of the elevation of the Deuteronilus shoreline. A later study of the highest-resolution MOLA data showed that *all ten* large channels that

---

* See Colour Plate 5 for the MOLA profiles of the shorelines.

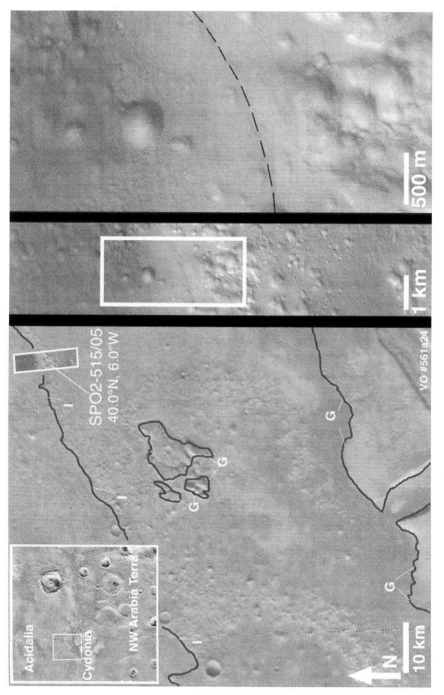

When Mars Global Surveyor inspected a 'shoreline' in the Acidalia Planitia–Cydonia Mensae region, the evidence was ambiguous.

drain north across the line of dichotomy terminate between the two putative shorelines, with even the largest channel having lost its morphology by the inner one. Head concluded from the termination of the Chryse channels that they drained into an ocean whose depth was 600 metres on average and 1,500 metres at its deepest point.

THE DILEMMA

With such strong morphological evidence for the previous existence of liquid water on the surface, it was inferred that Mars's early atmosphere must have been dense enough to sustain a hydrological cycle with both physical and chemical erosion of the surface, and that if an ocean existed for a prolonged time its bed must be lined with carbonate. The primary task of the thermal emission spectrometer on Mars Global Surveyor was to identify the sediments and evaporites deposited during this putative warm and wet era, but it found no trace of such minerals. However, it did find olivine, a silicate found in igneous rock, to be common, and since this mineral readily weathers in a warm and wet environment this discovery suggested that the planet must have been cold and dry since the olivine was formed, presumably billions of years ago! In an effort to resolve this dilemma, NASA decided to send another lander.

THE MARS EXPLORATION ROVERS

Shortly prior to the launch of Mars Pathfinder, the National Research Council had criticised the Sojourner rover for "restricted instrument complement and lack of significant mobility". But this tended to presume that the innovative entry, descent and landing system would work and it evaluated the mission in terms of scientific potential, whereas the objective was to demonstrate the landing system and, if this worked, gain some science as a bonus while testing autonomous robotics with the rover. Having proved these systems, it would be for later missions to deliver larger payloads for more far-ranging investigations. In April 1999 NASA conducted field trials of the rover for the Mars Sample Return mission. The rover's task would be to analyse rocks *in situ*, drill out cores, and put the most interesting samples in the return capsule. Known as FIDO, the development vehicle had greater autonomy than Sojourner, and a considerably wider radius of action. After the losses of Mars Climate Orbiter and Mars Polar Lander, the Mars Sample Return mission was put into abeyance, and it was decided to employ the airbag system to deliver a rover to conduct an independent mission. In fact, for added redundancy, it was decided to send two rovers. By this point, the great question was whether the Martian climate had once been warm and wet, and the strategy for determining this was to 'follow the water'. Mars Global Surveyor and Mars Odyssey were making remote-sensing surveys, and it was time to seek some ground truth. As robotic field geologists, the rovers were to have tools to investigate the physical characteristics, chemistry and

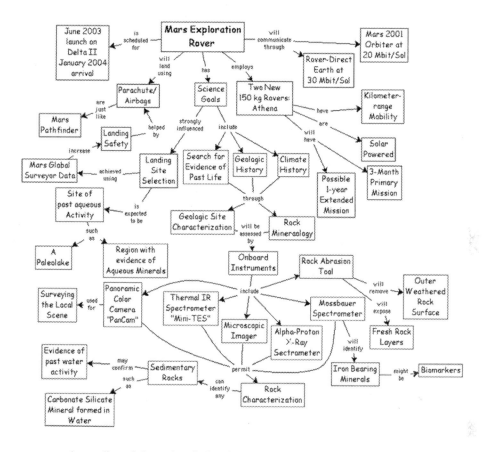

An outline of the rationale for the Mars Exploration Rover missions.

mineralogy of individual rocks in order to reveal whether they were either formed in or altered by liquid water.

At launch, the mass of the spacecraft was slightly over 1 tonne. The task of the cruise stage was to transport its payload to the vicinity of Mars and release it on course to land at the selected site. The payload comprised the forward heatshield, rear shell, parachute, airbag system, and folded tetrahedral lander, inside of which was the rover. Whereas Sojourner had relayed through the communications system on the Pathfinder lander, the new rover would be self-sufficient once it had rolled off its base. The body of the rover was a triangular box of a composite honeycomb material that contained the electronics. It was topped by the horizontal equipment deck on which were mounted a low-gain antenna, a UHF antenna, a high-gain disk antenna on a steerable mount, and a mast housing the main cameras. The modular solar array comprised the surface of the deck, a triangular flap at the rear, and two side flaps with a total area of 1.3 square metres, providing 900 watt-hours per sol at the start of the surface mission, supplemented by a pair of lithium-ion batteries. As in the case of Sojourner, the rover had six wheels, each having its own motor, on a

Technicians prepare one of the Mars Exploration Rover vehicles. The FIDO rover (top-left) was used to prepare for the mission. In its final form, the MER featured a robotic arm with a suite of instruments (top-right) to enable it to operate as a field geologist.

rocker-bogey suspension system incorporating joints to enable a wheel to roll over a rock that exceeded its own 20-centimetre diameter. Whereas the mass of Sojourner had been a mere 10 kilograms, the new rover was 174 kilograms. Its low centre of mass was designed to enable it to tolerate a tilt of 45 degrees but software would nominally intervene at 30 degrees.

The science package, known as Athena, was provided by an international team led by Steven Squyres of Cornell University. In order to overcome the perspective that had resulted from Sojourner's camera being located so close to the ground, the cameras on the mast had a viewpoint equivalent to that of a human field geologist. The panoramic stereoscopic camera had filters for 13 wavelengths in the visible and near infrared, and a resolution matching 20/20 eyesight. On the top deck was a chart for colour and greyscale calibrations and a sundial to enable the shadow of the pillar to measure the brightness of the ambient illumination. The monochrome navigation camera was of lower resolution but could record a wide horizon arc in fewer frames, and hence place less demand on the limited downlink capacity. The third instrument to exploit the mast was a miniaturised version of the thermal emission spectrometer on Mars Global Surveyor. Its electronics were housed in the main box, and the mast was used as a periscope. Its sensor produced a mosaic of 'false colour' circular spots that were later superimposed on an image to provide context. The longer the sensor 'stared', the greater was its signal-to-noise ratio. Each spot provided a spectrum in 167 wavelengths, from which temperatures and mineralogy could be inferred. It was to assist in the selection of rocks for individual sampling using the tools on a short robotic arm, developed by Alliance Spacesystems Incorporated in Pasadena, that was capable of motions involving five degrees of freedom.

The rover's arm had a microscopic imager with a resolution comparable to that of a hand lens. It was to assist in identifying rocks formed in water, features of volcanic and impact origin, and veins of minerals left by the presence of water in a rock. For soils, it would show the sizes and shapes of the grains, and provide insight into erosional processes. Like Sojourner, the new rover had an alpha-particle X-ray spectrometer to detect all the main rock-forming elements other than hydrogen. Because iron interacts strongly with liquid water, a Mössbauer spectrometer would investigate iron-bearing minerals. Since the rocks examined by the spectrometer on Sojourner had proved to be coated with weathered material, the new rover had a rock abrasion tool (RAT) developed by Honeybee Robotics in New York City. This could use a brush of stainless steel bristles to sweep dust off a rock and drive a pair of diamond teeth at speeds of up to 3,000 revolutions per minute to grind into the surface of a rock to reveal a circle of material 5 centimetres in diameter and several millimetres deep. As in the case of the Vikings, the rover also carried magnets. One set was on the RAT to collect dust from the grinding activities. A pair of magnets were mounted on the front of the vehicle, in positions where they could be inspected by the spectrometers. A third magnet was mounted on the top deck, in view of the mast-mounted cameras. Since the rover was to operate semi-autonomously, it had stereoscopic cameras facing forward and to the rear to enable it to avoid hazards. While the arm was operating, its activities would be monitored by the forward stereoscopic camera. The rover's orientation on the surface would be determined by using the navcam to locate the Sun, but its position would be calculated on Earth by correlating an odometer with triangulation using features on the horizon.

The landing sites had to be safe and suitable for the strategy of 'following the water'. The engineering requirements were that they be in the equatorial zone for illumination of the solar panels, at low elevation for the greatest parachute braking, and not so rocky as to inhibit driving. One site was selected for its morphological characteristics and the other on the basis of chemical remote-sensing from orbit.

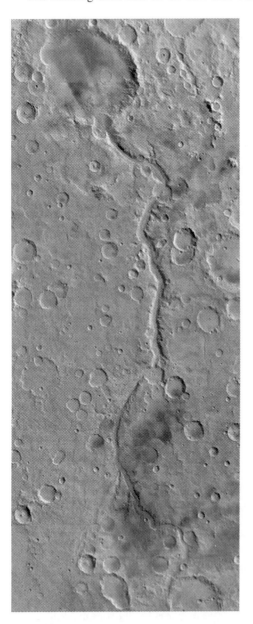

The southern rim of the crater Gusev has been breached by Ma'adim Vallis.

The first rover would be sent to Gusev, a 150-kilometre-diameter crater just south of the line of dichotomy. Ma'adim Vallis, a 900-kilometre-long channel running off the highlands, had breached the southern rim and, it seemed, formed a lake that left a deposit of sediment on the floor of the crater. The rover was to land on this putative lake bed to seek evidence of water action and, if possible, distinguish between sediments formed on the bed of a body of standing water, rocks formed in running water, and rocks formed in the absence of water but later modified by its presence. However, it was possible that such a lake bed might have been buried by volcanic activity, possibly from Apollinaris Patera to the north.

As for the site for the second rover, in 1998 the thermal emission spectrometer on Mars Global Surveyor indicated an exceptionally flat area of Meridiani Planum in the highlands just south of the line of dichotomy to be rich in grey hematite. The reddish hue of the planet derives from its surface having been oxidised. Hematite comes in two forms – red and grey – that are chemically the same but differ in the size of their crystals. The fine-grained red hematite that eroded from rock had been distributed planetwide by global dust storms, and was of no relevance to the search for evidence of past water. An environment involving liquid water is (usually, but not always) required to accumulate crystals of hematite into the large grains of grey hematite. The *in situ* investigation was to determine whether the hematite was present in layers of sediment in a lake, in veins resulting from the alteration of pre-existing rocks by a hydrothermal system, or, indeed, the result of another process not involving water.

7

Spirit

ARRIVAL

Spirit, the first of the Mars Exploration Rovers, was launched on 10 June 2003 and neared Mars on 3 January 2004. A regional dust storm on the far side of the planet in December had caused some concern. Infrared data from Mars Global Surveyor indicated that this dust had warmed the upper atmosphere on a *global* basis, and it was calculated that the diminished air density would reduce the efficacy of the first phase of the braking process. The interval from the deployment of the parachute to the radar firing the retrorockets would be cut by about 20 per cent, to 90 seconds. The event sequence nominally needed 100 seconds, so, as Brian Manning, leading the entry, descent and landing team, put it, "it looked tight on the worst case". The deployment of the parachute was to be triggered by the deceleration, as the canopy would be torn off if it was opened at too great a speed. A margin was bought by the expedient of reprogramming the computer to deploy the chute several seconds early, and accepting the risk of doing this at a slightly increased dynamic pressure.

Some 70 minutes prior to atmospheric entry, the cruise stage turned to point its payload's heat shield forward. Since this reduced the power from the solar array on the cruise stage, the lander switched to the rover's battery. Fifteen minutes out, explosive bolts released the lander, which was rotating axially at 2 revolutions per minute for stability.

As the loss of Beagle 2 had recently confirmed, trying to land on Mars was difficult. After the loss of Mars Polar Lander, NASA had ordered that future landers provide basic telemetry to indicate their progress. There was a low-gain antenna on the backshell to transmit a series of 10-second tones on achieving specific points in the descent sequence. The lander made contact with the atmosphere at a shallow angle at an altitude of 125 kilometres, travelling at 5.4 kilometres per second. Two minutes later, the deceleration peaked at 6 *g*, and the temperature of the heat shield peaked at 1,500°C. There was a possibility that the signal would be blocked by the plasma created around the vehicle, but this did not occur, and the tones continued. Having slowed to 400 metres per second, the 15-metre-diameter parachute deployed at an altitude of 7 kilometres. The engineers cheered when the deployment tone was received, but the most intensive phase of the descent was still to come, since 37

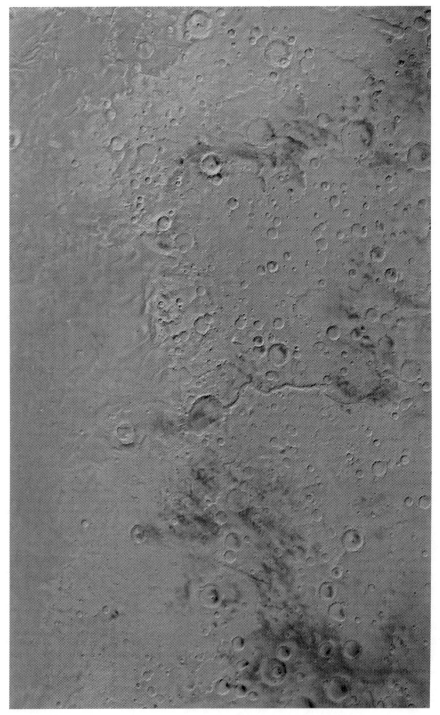

A Viking mosaic showing the channel of Ma'adim Vallis breaching the southern rim of Gusev crater, the line of dichotomy and Apollinaris Patera.

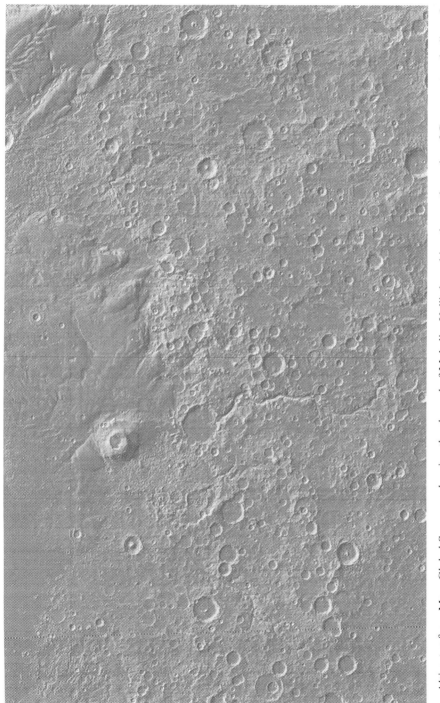

Altimetry from Mars Global Surveyor showing the channel of Ma'adim Vallis breaching the southern rim of Gusev crater, the line of dichotomy and Apollinaris Patera.

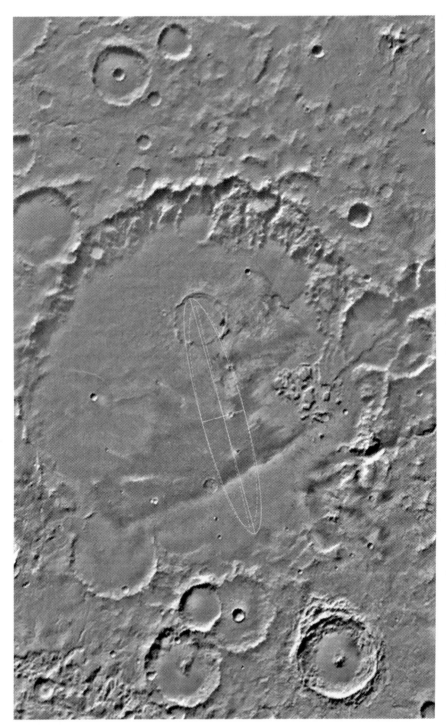

The preliminary landing ellipse for Spirit within Gusev crater.

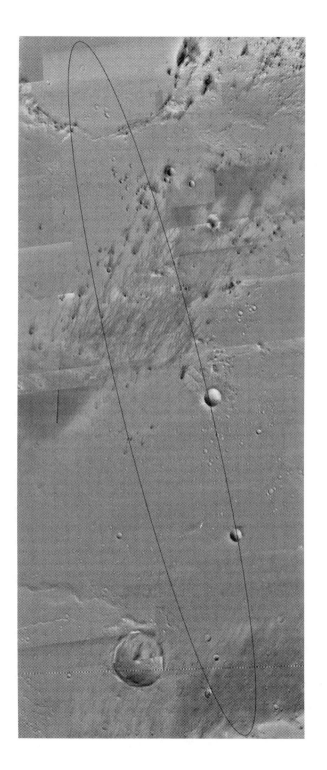

The final landing ellipse for Spirit within Gusev crater.

A 3-dimensional representation of Gusev crater, looking towards where the rim was breached by Ma'adim Vallis.

pyrotechnic devices had to fire in the proper sequence over the next two minutes. Twenty seconds later, the forward heat shield was released. After 10 seconds, the backshell started to unreel a tether, referred to as a bridle, to lower the lander, taking 10 seconds to reach its 20-metre length. At this point, the lander started to send UHF telemetry to Mars Global Surveyor, and the orbital spacecraft recorded this for later relay to the Earth. With 35 seconds remaining on the nominal profile, the radar altimeter was activated at an altitude of 2.4 kilometres, and several seconds later the downward-pointing camera of the Descent Imager Motion Estimation System (DIMES) snapped pictures at altitudes of 2, 1.75 and 1.5 kilometres, which were promptly analysed to estimate the horizontal velocity, in order to determine which of three transverse rockets on the backshell to fire to eliminate such horizontal velocity as might be caused by winds near the surface catching the parachute, or by the side-to-side swing of the lander on its bridle. At a height of 284 metres the airbags were inflated. There were six bags affixed to each of the lander's four faces, each a double-layered bladder to resist their puncture by rocks. As the transverse rockets fired two seconds later at a height of 100 metres, a trio of powerful retrorockets on the backshell were ignited and, a further three seconds later, with the lander halted at a

An artist's depiction of the key events in the entry, descent and landing sequence for a Mars Exploration Rover.

height of 9 metres, the bridle was cut and the still-firing retrorockets drew the backshell and parachute clear.

The tone to indicate that Spirit had made contact with the surface was received by JPL at 20:35 Pacific time and prompted a tremendous cheer from the engineers. The lander bounced to a height of 15 metres, then rolled across the surface. As it was impracticable to maintain the direct link to Earth during this time, it was 20:51 before the Canberra Deep Space Communications Complex was able to report that it was receiving "a very strong signal".

Working autonomously, at 20:56 Spirit began to deflate and retract its airbags, first the base petal and then the three side petals, completing the process at 21:30. As the lander could have come to rest in any orientation, the electrically driven hinges on the side petals had sufficient torque to flip the lander onto its base, on which the rover was mounted, but it had actually come to rest the right way up. As the petals were opened, mats of a similar fabric to the airbags were draped across the struts, hinges and wire harnesses to prevent these impeding the passage of the rover when it later rolled off the lander. Once the petals had deployed, the rover unfolded its solar panels to begin to recharge its batteries. At the landing site it was early in the afternoon on what would be defined as sol 0. When Mars Odyssey flew overhead several hours later, the lander uplinked 24 megabits of data by UHF, and when this began to stream in at 23:30 it was greeted by applause as this confirmed that the rover had survived the landing and was operating autonomously. There was more applause when it was realised that this included the first monochrome imagery. A view looking forward showed that although there was a section of airbag bunched up beyond the ramp, there were no rocks blocking the exit path. Ten minutes later the individual frames had been made into a panoramic view, showing a level plain with rocks of various sizes. "Welcome to Gusev crater," someone called out. "It's not supposed to go this well," mused one of the flight controllers, reflecting on the simulations in which there had always been 'issues' to be addressed.

The DIMES images had been downloaded during the Mars Odyssey relay, and indicated that the landing was 10 kilometres beyond the centre of the ellipse, in an area streaked with tracks from dust devils. This was seen as a bonus since it raised the prospect of the wind having cleaned the rocks by removing the surficial fines.

"The rock population is close to ideal," reported lead scientist Steven Squyres of Cornell University at the press conference several hours later. "We're going to be able to motor around." Compared to the previous sites, this was remarkably flat, which made sense if it was a lakebed. If it had been as undulatory and littered with rocks as the Pathfinder site, the rover's radius of action would have been severely restricted. There were, however, limits to what he could say at this stage. "We don't know what kind of geological material we have landed on – don't have a clue. I don't know what kind of rocks they are. We haven't seen any pancam yet, so we've got no colour, we've got no infrared. That's going to come in the days ahead. So I don't know if these are sedimentary rocks or if they're lava that's been deposited over it, but if you asked me ahead of time what's a dry lakebed on Mars going to look like, I'd have said a lot like this!"

As the initial imagery from the mast-mounted navcam had been compressed by a

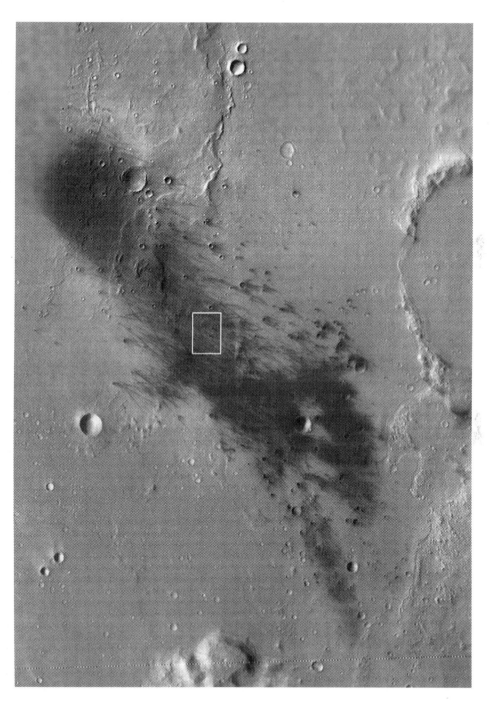

This Mars Odyssey image shows the dust-devil streaked terrain on which Spirit landed.

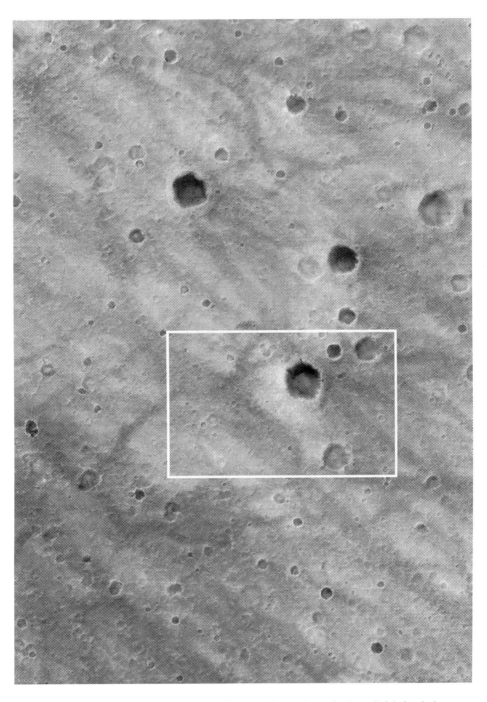

This Mars Global Surveyor image shows a closer view of where Spirit landed.

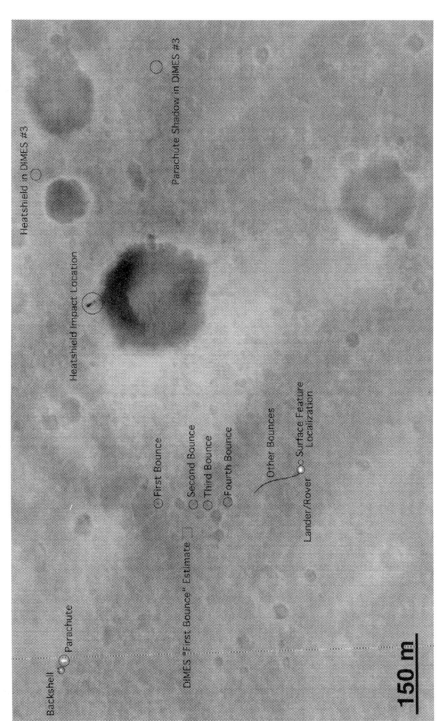

An analysis by Malin Space Science Systems of the Spirit landing site.

factor of 16 to enable it to be downlinked via the first UHF relay session, it was of modest resolution, but there were several intriguing features nearby, including some depressions with rock exposed around their margins. One patch, several metres to the north, across which the lander had rolled, was named Sleepy Hollow and was probably a secondary impact. Squyres noted that while he was eager for a closer look, the fact that it seemed to contain "very smooth-looking soil" meant that it might be a "rover trap".

SURVEYING THE SITE

Two hours after sunrise, having been on the surface for 16 hours, Spirit awoke for sol 1. It used its navcam to locate the Sun, calculated the offset for Earth, raised its high-gain antenna and slewed this around to establish direct communications. If it had failed to link up, it would have continued with its preprogrammed sequence of activities for the full 90-sol duration of the primary mission. At 14:30 local time, after completing basic tests of its scientific instruments, the rover activated its pancam to record a high-resolution colour view of the horizon directly ahead, and this was returned overnight via satellite relays. Software written by Jascha Sohl-Dickstein processed imagery taken using different filters and made a multispectral analysis, and software developed by Jonathan Joseph created mosaics. Each frame was 1,000 pixels square, and this first set was 3 frames wide and 4 frames high. As the high-gain antenna was operating at its medium bit-rate at this point, the frames had been compressed slightly, but the resolution was nevertheless astonishing.

"This is Mars like you have never seen it before!" proclaimed Squyres. By operating the UHF transmitter at its maximum bit-rate, the rover could transmit 50 megabits during one relay pass. Allowing for the fact that this bandwidth could not be solely devoted to the pancam, the 360-degree panorama had been divided into 45-degree segments. It would therefore require most of the 10 days or so assigned to the rover's preparations for driving off its lander to send this pan and the corresponding mini-TES panorama. In effect, these initial imaging tasks were what an immobile lander would have done. They formed a major objective of the science programme, and would, as Squyres put it, "provide the framework for where our explorations will take us in the days immediately after egress". While it was too early to draw any conclusions from the rocks, he pointed out that their distribution was "remarkably different from anything we've seen on Mars". In comparison to either the Viking sites or the Pathfinder site, "without question, there are far fewer big rocks". The surfaces of the rocks were remarkably smooth, with shapes ranging from very rounded to quite angular, suggesting that they were very hard, made of fine-grained material, and had been polished by aeolian erosion over a very long interval.

Michael Malin noted that there were a significant number of fractured rocks. If thrown in as ejecta, these could have been broken on striking the ground; if local, they might have been fractured by environmental effects. Shock would make sharp fractures, but weathering could produce a range of effects. "My work in Antarctica has led me to lots of rocks split by freeze-and-thaw and other physical weathering

The first high-resolution pancam image returned by Spirit heightened suspicion that the floor of Gusev crater was a dried up lake bed.

processes – almost all of which require water in a liquid state. The fact that I see more fractured rocks at this site than in all the other sites combined suggests to me that at some time in the history of *this* surface, water was involved in the physical breakdown of the rocks."

The pancam view showed detail that had not been evident to the navcam. One intriguing observation was that as the airbags had deflated and been drawn across the surface, they seemed to have forced pebbles down into the soil, which resembled layers of cohesive material. There were also 'scratches', where pebbles had been dragged, peeling and folding the soil. "The way in which this surface has responded to the retraction – the dragging of the airbags across it – is bizarre," said Squyres. When pressed, he said, "It looks like mud, but it can't be." In response to prompts by reporters that what looked like mud must surely be mud, Malin pointed out that "very, very find-grained material behaves in strange ways, including like liquid". Once the pancam had returned a less compressed image for greater resolution, one small curled up section about 10 centimetres across and a few millimetres thick was called the Magic Carpet. John Grotzinger, a geologist at the Massachusetts Institute of Technology, observed: "At low resolution it looked like a fluid mud, but in higher resolution it doesn't have the mechanical behaviour of mud." He was intrigued that this piece of duricrust – or a caliche of loose rock and dust cemented together with salts – "doesn't exhibit the brittle deformation that we saw at the Pathfinder site" where airbags had been used for the first time. David DesMarais, an astrobiologist at the Ames Research Center in California, speculated that the Magic Carpet might be the result of concentrated brine in the very fine-grained material, because a sulphate brine could remain fluid at subzero temperatures and evaporate only very slowly, but Malin countered that the very fine particles were more probably bound electrostatically.

What had looked plausibly like a dry lakebed in the navcam imagery started to seem less so as the higher resolution pancam worked its way around the horizon. The arc downloaded on sol 5 viewed northward. Raymond Arvidson, a geologist at Washington University in St Louis, noted that the rolling rock-strewn landscape was different to the smooth windswept lakebed envisaged from orbital imagery. "A lakebed is typically flat, with very fine-grained sediments," he pointed out. "That's *not* what we're looking at. It's a surface strewn with rocks – a surface that probably has a number of secondary craters that excavated rock. It's not a primary depositional surface of a lakebed such as you'd see on your way, for example, between Los Angeles and Las Vegas."

"We've talked about Gusev being a dry lakebed for so long, there's a danger of becoming trapped into believing only in this possibility," warned John Grant of the National Air and Space Museum in Washington, DC. But there were alternative hypotheses. "To be trite, you can't judge a book by its cover," said Jeff Moersch of the University of Tennessee at a press conference. He presented two pictures taken on Earth that resembled Spirit's location. At first sight, the shapes of the rocks at the terrestrial sites, their sizes and their distributions, could be interpreted to imply that they were formed by the same process, but when the sites had been examined in detail they had been revealed to have formed in different ways. In the case of the

Haughton Impact Structure – an impact crater in the Canadian Arctic that formed 23 million years ago – a lake had pooled on its floor. However, the manner in which some of the rocks were scratched indicated that they had been transported by glacial activity, and were unrelated to the site. On the other hand, the Saf Saf area of Egypt, which was the second site, was not a lakebed. "When I saw the pictures of Gusev," Grant said, "I caught myself saying 'I know this place', because it looks like places I've seen on Earth, but we must be very cautious because we know we have *never* seen this place before. If we only had *pictures* of the site, we scientists would argue for decades, and maybe never know for certain how this place formed." However, Spirit was equipped for the kind of field geology that *would* resolve the issue. When the pancam mosaic extended around to the northeast, it revealed the rim of the largest of the three craters seen in the DIMES imagery. It was 275 metres away and 200 metres in diameter. By the manner in which an impact makes a crater, the material from the deepest part of the excavation is dumped on the rim, which in this case rose 4 metres above the surrounding plain. If the lakebed had been buried by a lava flow, the best chance of sampling the sedimentary rock might be in the ejecta that surrounded this crater. To the southeast, some 2.3 kilometres away, was a group of hills rising about 100 metres above the plain which, as Matt Golombek put it, were "the nearest 'different' things from where we are now". If the hills had been submerged by the lake, then by projecting above whatever had buried the lakebed on the open plain they might hold the key to the mission.

As the mini-TES panorama built up, it was superimposed on the corresponding pancam view to interpret the thermal and multispectral data. Sand heats up more rapidly than rocks after sunrise, and cools more rapidly after sunset. By identifying dust, mini-TES would enable Spirit to avoid becoming 'bogged down' in dust. The uniform temperature across the depression of Sleepy Hollow confirmed that it was dusty, and hence should be avoided. The multispectral data indicated that the dark rocks were of a basaltic composition. There were rocks of a lighter tone that might be sedimentary, but this would not be able to be investigated until the rover made a closer examination.

THE TRAVERSE PLAN

The nominal plan for the 90-sol primary mission had been to undertake a series of sampling activities ranging out 600 metres from the lander, but the inviting terrain had prompted calls to go further. There had been considerable debate because, as Arvidson told journalists, "We have an excellent science team, but it's like herding cats; they're all trained to be totally independent and vocal!"

"What we have is a vast plain," said Squyres when the complete panorama was presented on 12 January. "We'll soon know what the material we're perched on is like. But what lies beneath that? And what's higher up? Those are the questions we'd like to answer by traversing. So we want to go some place where there's a big hole in the ground." It had been decided that after sampling in the vicinity of its lander, Spirit would begin its exploration by driving to the crater to the northeast. "I don't know what we'll do when we get there, but as we approach we'll enter the ejecta blanket. We'll be careful as we close in, as no one has ever driven up to a Martian crater before! We might drive right up to the lip of the crater, it depends on trafficability. It will give us a 'window' into the surface – certainly as deep into Mars as we'll see on this mission." The presence of layering in the interior wall might indicate the thickness of the material overlaying the lakebed. The subdued profile of the crater indicated that it was a fairly old secondary crater. "After that," continued Squyres, "we'll swing southeast for the hills." These were the nearest 'etched' terrain on the geological map compiled from overhead imagery. "I can't tell you we're *going to reach* the hills – we're going to go *towards* them." As Spirit approached, the chance of seeing material on the plain that originated on the hills would increase and, with the improving imaging resolution, it would be possible to look for outcrops of rock on their flanks. And if the mission could be extended, Spirit might actually be able to cross the 'contact' between the two morphological units and conduct an *in-situ* investigation of the lower slopes.

EGRESS

On sol 4 the motor on the base petal was activated to draw in the section of airbag that was obstructing the primary exit path. This helped, but further retraction was ordered on sol 5, at which time the appropriate side petal was tilted up 20 degrees to allow more space prior to activating the retraction motor. Reporting that this "lift and tuck" had not improved the situation, mission manager Matt Wallace told reporters that another attempt would be made on sol 6, after which the petal would be driven down. "If it is successful, we'll most likely egress off the front of the lander; if not, then we have several options, one of which is to turn to the right about 120 degrees and drive down the ramp in that direction." On failing to make further progress, it was concluded that some of the lanyards must have been weakened by the landing, and snapped when the deflated airbags were retracted. A turn-in-place would pose no difficulty, as the deck was almost horizontal.

The extent to which the landing bags had retracted is clearly seen in this overhead projection of navcam imagery.

Once the lander had opened its petals, Spirit had unfolded the segments of its solar panels and raised its mast. It was, however, 'hunkered down', and latched to the deck of the base petal. Having it 'stand up' was nominally a two-day activity. The first part of this process was performed on sol 6. A screw-jack mechanism on the lander lifted the body of the rover 15 centimetres above the deck in order to unfold the rocker-bogeys. The front wheels, which were folded in front of the rover, were

An artist's depiction of a Mars Exploration Rover about to roll off its base, a view from Spirit's forward hazard camera immediately prior to making the move, and a view from the rear hazard camera confirming that it had done so.

then swung around into position, and the jack was lowered until latches locked the undercarriage. The second part of the process began with the retraction of the jack into the lander's deck to avoid any obstruction, then the rear wheels, which were stowed against the middle wheels, were deployed rearward into their operating position. Next, the latches that held the middle wheels on the lander were released. Finally, the arm was unlatched and stowed in its driving position. Chris Voorhees, the engineer who designed these mechanisms, told reporters, "It's been a very, very, very exciting couple of days for the spacecraft team." The process had involved 12 pyrotechnic devices, nine motorised mechanisms, and six structural latches. The final step, on Saturday, 10 January, was to sever the umbilical between the rover and the lander, by way of which the petals and airbag retraction motors had been commanded. The stand-up had taken so long owing to the requirement to confirm each step before advancing to the next step.

It was decided to have Spirit rotate right through 115 degrees on sol 10, but as a preliminary, in order to gain manoeuvring room, it was to reverse 25 centimetres. After rotating 45 degrees it was to pause and take a picture of a blind spot that the solar panel had masked, in order to verify that there was no hazard at that location, then rotate a further 45 degrees and make another check prior to finishing the turn. If all went well, the rover would depart the lander on sol 12. "The most dangerous driving we're going to do, is the first several metres," flight director Chris Lewicki warned reporters, "because our lander is the most dangerous obstacle." The rocks beyond were about 20 centimetres in size, but posed no threat since the suspension system would enable the rover to run over them. On command, Spirit drove for 78 seconds at 4 centimetres per second. An hour later, having realigned its high-gain antenna, it transmitted a picture taken by its rear hazard camera, showing that it had halted with its rear wheels 1 metre clear. As Joel Krajewski, the chief engineer for the lander's part of the surface mission, put it, Spirit had "completed its landing".

"We have six wheels in the dirt!" announced JPL director Charles Elachi at the 06:00 press conference. Kevin Burke, leader of the egress team, was exhausted. "I have to tell you that being the person who has the last piece of hardware between sticking on the lander and being on the surface of Mars is very, very stressful." He was looking forward to having a fortnight's rest before Opportunity's arrival.

TO WORK

Spirit was to spend several sols at the egress site, making its initial observations. "We want to characterise the geological diversity," reported Squyres. "That means going and finding the unusual rocks, finding the unusual soils, finding the things that are *not* characteristic of the typical stuff. Before we can do that, we've got to understand the typical stuff, so the first thing that we'll do is slap the arm down to sample the soil." This would be the first use of the instruments and the initial pace would be slow and deliberate because, as John Callas, the science manager, put it, "we want to get it right". One engineering test, for example, would be to determine how accurately the arm could position its instruments on or above the surface. The egress site was not ideal for soil characterisation because it also contained pebbles, but it would suffice for testing the instruments.

On sol 13 the arm positioned the microscope to inspect the soil, one picture was taken, then the arm was repositioned slightly to facilitate stereoscopic analysis. Ken Herkenhoff of the US Geological Survey, the leader of the microscope team, noted that the soil was "a conglomeration of dust particles" with the consistency of "clumpy cocoa powder". After the Mössbauer spectrometer was emplaced against the ground for a 4-hour integration on sol 14, the microscope examined the surface and found that the instrument had left almost no imprint. "What I suspected", said Squyres in surprise, "was that the grains were held together by electrostatic forces, and what I thought would happen was that the instrument would scrunch this stuff, and the grains would collapse and flatten out like talcum powder. It didn't happen. Nothing collapsed. So what is holding these grains together?" Finally, the APXS was deployed just above the surface to integrate for 20 hours overnight. Goestar Klingelhöfer from Johannes Gutenberg University in Mainz, Germany, leading the Mössbauer team, reported that his spectrometer had detected the presence of three different iron-bearing minerals, the most abundant of which was olivine, a silicate found in igneous rock which, on Earth, soon weathers to clays and iron oxides in the presence of water. Orbital instruments had shown olivine to be widespread as a constituent of the dust, but its high abundance in the soil was surprising. Johannes Brückner of the Max Planck Institute for Chemistry in Mainz, a member of the APXS team, reported that this instrument had found high concentrations of silicon and iron, with lesser amounts of calcium, sulphur, chlorine and nickel. The levels of chlorine and sulphur were characteristic of the soils at the previous sites. "There might be sulphates and chlorides binding the little particles of soil," said Squyres. As

COLOUR SECTION

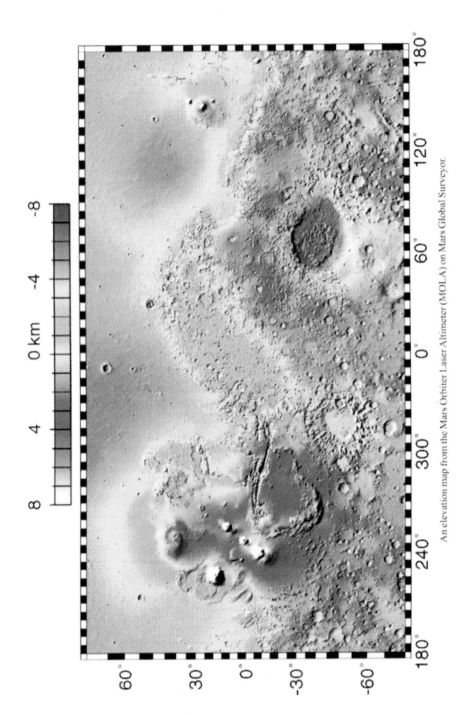

An elevation map from the Mars Orbiter Laser Altimeter (MOLA) on Mars Global Surveyor.

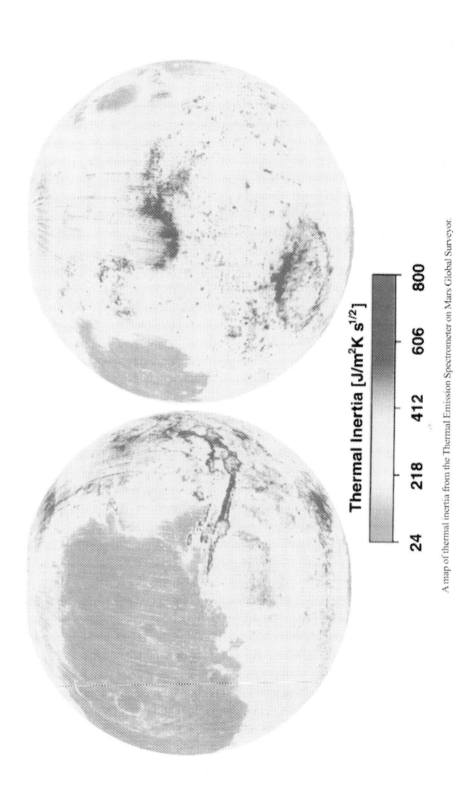

Thermal Inertia [J/m²K s^(1/2)]

24   218   412   606   800

A map of thermal inertia from the Thermal Emission Spectrometer on Mars Global Surveyor.

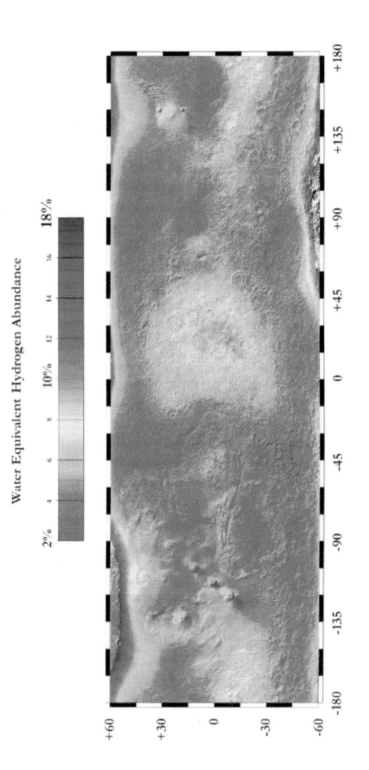

Water Equivalent Hydrogen Abundance

2%   4   6   8   10%   12   14   16   18%

A map from the neutron spectrometer on Mars Odyssey in which the presence of hydrogen is interpreted as water abundance in the topmost metre of the surface.

A view by Mars Pathfinder of its rover Sojourner inspecting a nearby rock, with Twin Peaks on the horizon.

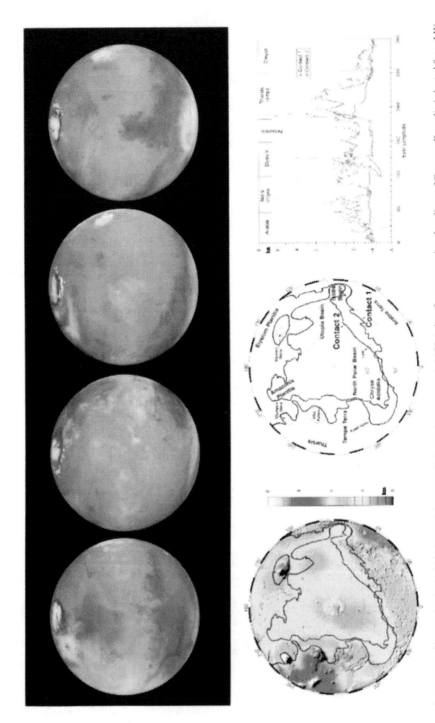

A sequence of images by the Hubble Space Telescope of Mars as it rotates (top). MOLA data for two putative shorelines of Oceanus Borealis (adapted from J.W. Head et al, *Science* vol 286, 1999).

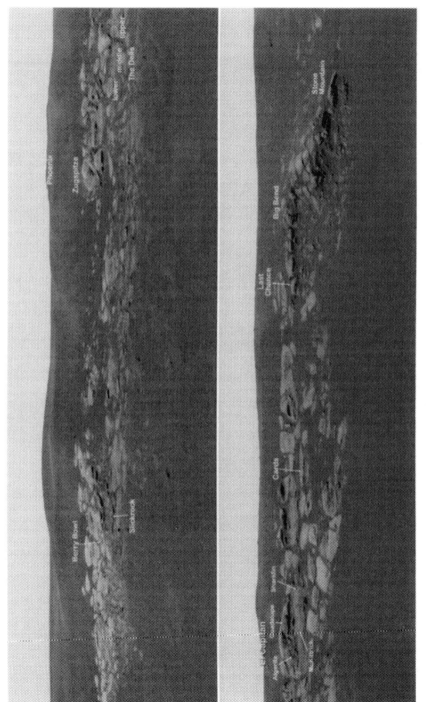

An annotated view of the outcrop in Eagle crater.

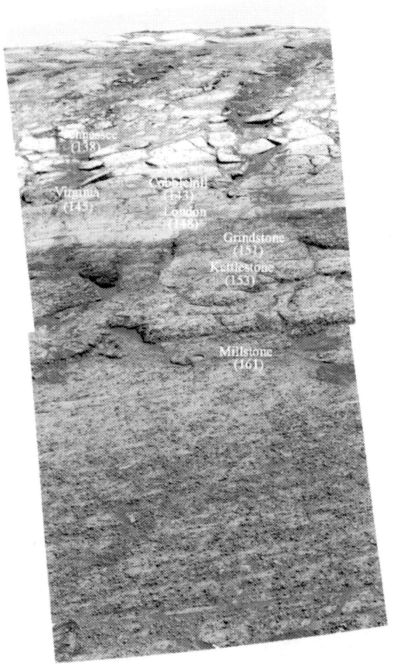

A false-colour view highlighting the holes that Opportunity drilled in the Karatepe exposure in the wall of Endurance crater.

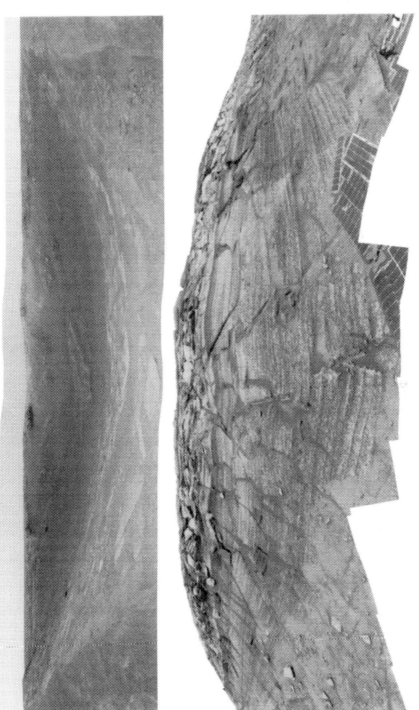

The interior of Endurance from the crater's western rim, showing the Burns Cliff on the right, and (bottom) the cliff viewed from a vantage point at its base.

A section of the first pancam mosaic returned by Spirit showing the floor of Gusev crater.

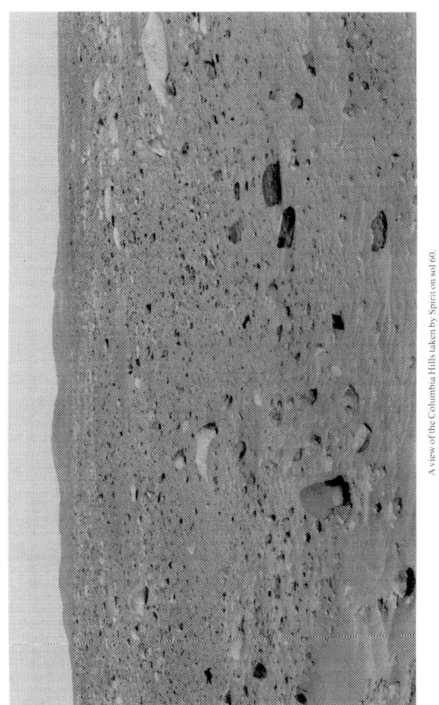

A view of the Columbia Hills taken by Spirit on sol 60.

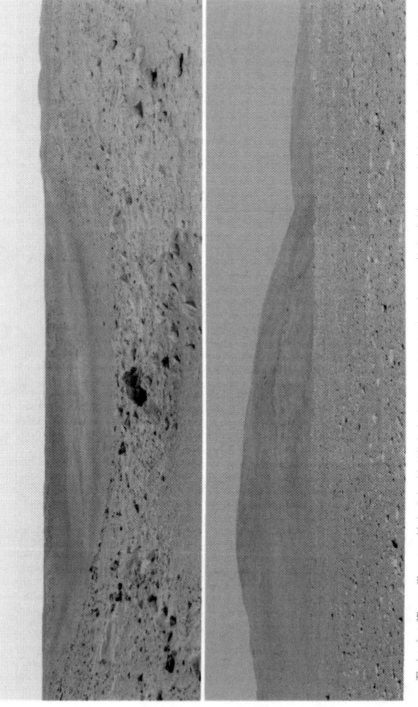

The interior of Bonneville crater with the Columbia Hills in the distance (top), and a view taken on sol 149 as Spirit approached the West Spur of Husband Hill.

500 m

A geological map of Spirit's route across the floor of Gusev crater to the Columbia Hills.

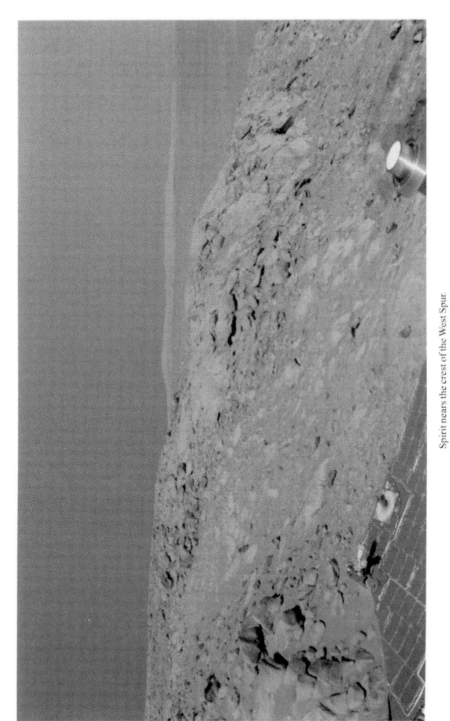

Spirit nears the crest of the West Spur.

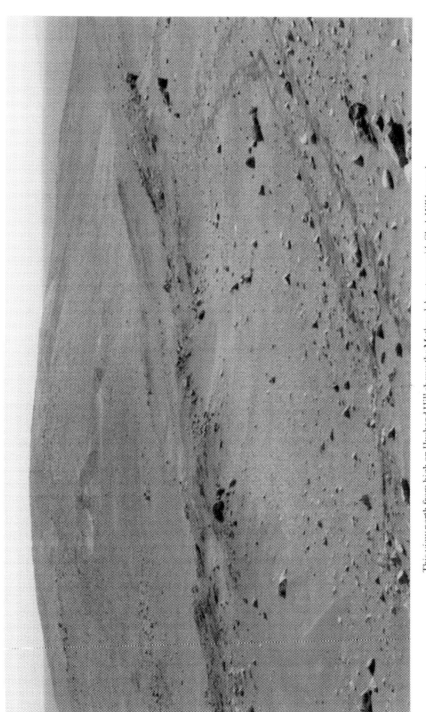

This view north from high on Husband Hill shows the Methuselah outcrop with Clark Hill beyond.

On sol 489 Spirit recorded this sunset from high on Husband Hill.

always, there were competing hypotheses. "Such salts", he noted, "could have been left behind by evaporating water, or been delivered in volcanic deposits." And, of course, the fines may have originated anywhere on the planet and been blown in by dust storms. Albert Yen, a member of the JPL team, suggested that the nickel was derived from meteoritic debris and, if so, the ferric iron compounds might have been produced by reactions involving water vapour in the atmosphere rather than liquid water. The indications from this initial sample were that the surficial material had not been wet. If there had been a lake and this left sediments behind, these must be buried. "Mars isn't going to give up its secrets easily," Squyres ruefully observed. The olivine was "an intriguing puzzle". It was readily weathered, but was not itself a weathering product. "We had expected weathering material like iron oxides, and we haven't seen these yet", Klingelhöfer said. "One possibility", Squyres speculated, "is that rather than being the result of chemical weathering, the soil is simply finely ground-up lava." The fact that there were pebbles at the sample site held out the possibility that the olivine was in the pebbles rather than the soil.

The next task was to inspect one of the nearby rocks. At the egress point, Spirit was facing northwest. There were two rocks near Sleepy Hollow named Sushi and Sashimi, and another off to the left, named the Pyramid on account of its shape. It was decided to examine the Pyramid, since it appeared to be free of dust and had a flat face that would readily facilitate grinding by the RAT. On Sunday, 18 January, the rover turned left through an angle of 40 degrees in a series of short arcs, made a turn-in-place to face the football-sized rock, now renamed Adirondack, and then advanced 2 metres into a position 30 centimetres from it. Although the vehicle was in motion for a total of only 2 minutes, the manoeuvre lasted 30 minutes. "The drive was designed for two purposes," said Eddie Tunstel, who had choreographed the activity, "one of which was to get to the rock. From a mobility engineer's point of view, this drive was to test *how* to drive on this new surface – as opposed to in the sandbox at JPL." That is, it was a calibration test to estimate how much the wheels slipped in the soil, prior to ordering much longer drives. The working hypothesis, prior to sampling, was that Adirondack was of volcanic origin. The first act in the sampling process would be to advance the arm with the RAT facing the rock, and wait until contact sensors indicated that the device was hard against the surface of the rock. The RAT would then be rotated out of the way to enable the microscope and spectrometers to inspect the surface prior to the grinding operation, and as this operation progressed the Mössbauer was left to integrate overnight on sol 17.

CRISIS

On Wednesday, 21 January, several hours after sunrise on sol 18 for Spirit, the daily command uplink by Canberra was marred by a thunderstorm, and although the rover acknowledged receipt it failed to start the scheduled high-rate direct link to Earth an hour later, and two hours after that failed to establish a UHF relay with Mars Odyssey, but a few hours later Mars Global Surveyor received a very brief transmission. Seemingly, the uplink had been corrupted by electrical interference from the thunderstorm and, despite having been programmed to reject erroneous commands, the rover's computer must somehow have 'crashed'. But the following day project manager Peter Theisinger reported that the issue was "a very serious anomaly on the vehicle". His deputy, Richard Cook, added that when Mars Global Surveyor had briefly received a signal, Spirit "was only sending a random pattern of zeroes and ones – effectively, what that means is that the transmitter was on but the computer wasn't sending information". This was ominous since, as Theisinger pointed out, "there is no single fault that explains all the observables". Nevertheless, several hours later he said that, on the assumption that Spirit had entered a safe mode and was listening at 10 bits per second, a signal had been sent to command it to issue a tone – and a tone had been received. One good thing, he pointed out, was that in contrast to a spacecraft in space that could all too readily lose its lock on the Sun and drain its batteries, the rover was in a stable orientation on the surface and was guaranteed solar power, which meant that there was *time* to investigate and resolve the problem. The first step would be to retrieve engineering data from the stricken rover.

Early on Friday, at 03:00 at JPL and several hours after sunrise at Gusev, the Deep Space Communications Complex near Madrid in Spain sent a low-rate signal to Spirit commanding it to transmit engineering data directly to Earth, prompting a 10-minute transmission at 10 bits per second, which was corrupt. But then, several hours later, Spirit sent 20 minutes' worth at 120 bits per second which established that, as Theisinger put it, the software was "not behaving normally". The processor was evidently in a 'reset loop' in which it woke up and loaded the flight software, then detected a condition that prompted another reset. The alarming thing was that the cause of the reset was not always the same. Nevertheless, this was a start in the diagnostic process. Despite having been commanded to sleep, the rover remained awake all day and, surprisingly, in the late afternoon Mars Odyssey reported that it had received 73 megabits at 128 kilobits per second by UHF with engineering data from earlier that day. Having something significant to work with, an anomaly team was formed to focus exclusively on diagnosing and overcoming Spirit's problem. Summing up, Theisinger said that he was optimistic that Spirit would be able to be restored to operation, although probably somewhat degraded, but doing so would take "many days, perhaps a couple of weeks".

By Saturday the working hypothesis was that the issue involved the computer's 'flash memory', as a fault in accessing this could readily give rise to a wide range of anomalous behaviours. The computer had three kinds of memory: there was a small amount of electrically programmable memory (EPROM) containing the core of its operating system, and 128 megabytes of random access memory (RAM) for real-time storage, but because this was 'volatile' – it lost its data when the power was switched off as the rover went to sleep – there was also 256 megabytes of flash memory (of the type used in commerical digital cameras) that could retain its content without power. The flash memory was organised as a file system (such as on a home computer) and held software and the data awaiting transmission to Earth. Every morning, when Spirit woke up, it would load into RAM the software that it would require for that day's operations, and before shutting down it would archive the data in its RAM that required to be preserved. Fortunately, there was an alternative operating mode. If instructed, the computer would start up without the flash memory, and operate in a limited capacity using only its volatile storage. When so instructed later that day, Spirit started up in a healthy state and for 1 hour transmitted engineering data at 120 bits per second, then obediently went to sleep in

This image from the forward hazard camera confirmed that, despite the problem that temporarily disabled Spirit, the arm was still in place against Adirondack.

order to recharge its batteries. "So our working hypothesis has been confirmed," reported Theisinger. "To the best of our estimation, the fault protection worked as designed. It just took us a long time to figure out what was going on." The next job was to establish a high-rate link and try to retrieve key files from the flash memory to further characterise the problem. On Monday, mission manager Jennifer Trosper was able to announce, "I think we've just found an issue with the number of files that had accumulated over the course of the cruise and the 18 sols on the surface." The amount of space required to manage the flash memory had exceeded that available in RAM. The solution would require deleting some of the files in the flash memory, but this had to be done carefully because some of the files contained data that had not yet been downloaded.

With the reintroduction of the high-gain antenna on Wednesday, it became possible to increase the rate at which the flash memory could be downloaded. In addition, Spirit returned its first image since the onset of the problem – a view from the forward hazard camera showing the arm with the Mössbauer emplaced against Adirondack just where it would have been after its overnight integration on sol 17. This was good news, because it meant that Spirit had not undertaken any physical movements while it was in trouble. The files containing the Mössbauer and APXS data of Adirondack were retrieved the next day, together with microscope pictures of its surface. The presence of olivine, pyroxene and magnetite indicated it to be of basaltic composition. "If you were a geologist on Mars and whacked that rock with a hammer, it would ring," Arvidson told journalists.

To mark the anniversary of the loss of the STS-107 crew on 1 February 2003, NASA named the site the Columbia Memorial

On the rear of Spirit's high-gain antenna was a memorial to the Columbia crew.

Station. There was a plaque on the rear of the 15-centimetre-diameter disk of the high-gain antenna.

BACK TO WORK

After the cruise files were deleted from the flash memory on Friday, sol 28, Spirit was restarted in its normal mode. "Spirit appears to be working just fine," reported Glenn Reeves, the chief engineer for the flight software. The remainder of the high-priority files were downloaded over the weekend, and on sol 31 the partition of the flash memory allocated to the file system was reformatted. On sol 33, Thursday, 5 February, Spirit resumed its interrupted study of Adirondack by using its RAT for five minutes to brush away air-delivered dust. The dark circular patch this revealed came as a surprise, as the rock had been selected for appearing to be fairly free of dust. This confirmed the need to clean a rock prior to measuring its chemistry and mineralogy. Over an interval of three hours the following day, the grinder drilled a hole 2.7 millimetres deep, and the brush then swept the dust out of the hole. "This has been a great day for robotics in planetary exploration," pointed out Stephen Gorevan of Honeybee Robotics, which had built the tool. The next day, first the Mössbauer and then the APXS were inserted into the cavity to analyse the exposed rock, and then the microscope took a look. "It's a beautifully cut, almost polished surface", said Squyres. The results confirmed the early assessment. "Adirondack", concluded Arvidson, "isn't the kind of 'smoking gun' evidence we're looking for in terms of climate history."

There was another rock nearby, named White Boat, whose light-tone raised the prospect of it being of sedimentary origin. On sol 36 Spirit made a 6.4-metre drive around the south of the lander to examine this rock, but when it was shown by the mini-TES to be a dusty companion to Adirondack the decision was made to ignore it and go to the crater, which had been named Bonneville. "We'll start out slow," said mission manager Jim Erickson, "and build up as we gain experience driving." For the first leg of 21.2 metres on sol 37, the rover was told where to go. Once its mobility was confirmed, the autonomous navigation software would be activated in order to pick up the pace. Allowing for pauses *en route* to investigate targets of opportunity, it was hoped to cover the distance in about 18 sols. Ron Baalke of the public affairs office at JPL reflected upon the irony: "We did our darndest to avoid landing in a crater for hazard reasons, but now that we're safely on the ground the crater has become the primary target."

TO BONNEVILLE

On awakening on sol 38, Spirit remained in place in order to examine a patch of rippled soil. The following day it drove 24.4 metres over a period of 2 hours and 48 minutes and halted in front of a group of rocks named the Stone Council, which it inspected using its arm tools on sol 40. One of these rocks, named Mimi, looked

A view of Adirondack prior to inspection (top-left), the result of brushing it (top-right, with microscope image), and after drilling it (bottom set, with microscope image).

Two views of the flakey rock Mimi.

different to any rock so far seen at Gusev. Its flakey appearance prompted several hypotheses: one, it was a rock that had been subjected to intense shock in an impact; alternatively, it was a section of dune material that had been cemented into loosely adhering layers, which was interesting because cementation can involve the action of water. "The science community like what they're seeing, and want a closer look," said Matt Wallace. Spirit shuffled into position on sol 41 in order to examine Mimi using the arm tools, which it did the next day; the rock proved to be basaltic.

The drive to Bonneville resumed on sol 43, this time in two legs. After driving 19 metres in the morning to a point specified by Earth, the rover took more stereoscopic imagery to update its map of the hazards and in the afternoon drove a further 8.5 metres on its own to a site named Ramp Flats. On awakening on sol 44, Spirit lowered its arm to examine the soil using its microscope and, while the Mössbauer was at work, the pancam studied nearby rocks. On sol 45 Spirit drove 22.7 metres to the next objective, which in this case was Laguna Hollow, a shallow depression similar to Sleepy Hollow. In pursuing its radial approach to Bonneville, Spirit was seeking systematic differences in the soil and the types and sizes of the rocks in order to characterise the ejecta blanket. On the morning of sol 46, Spirit 'scuffed' the dust using its front wheels, then reversed in order to allow the mini-TES a clear view of the disturbed material. In the afternoon it used the microscope, set the Mössbauer to work and went to sleep, then awoke at 02:00 to swap to the APXS. On sol 47 Spirit spent two hours manoeuvring in place to enable its front wheels to excavate a trench to a depth of 7 centimetres. The cohesive nature of the soil made the walls resistant to collapse, which prompted further speculation that it might be moistened by brine. After reversing to give the mini-TES a view of the trench, Spirit returned to make a microscopic study of the exposed material and use the spectrometers. The reason for studying the windblown sediments which had accumulated in Laguna Hollow was to seek insight into atmospheric processes.

In terms of duration, Spirit was now half way through its primary mission, but everyone was confident that it would be able to continue well beyond the 90 sols defined as the mission-success criterion. Richard Cook said that things had "gone better than we expected". In the case of Sojourner, a number of issues had limited its effective work to, on average, one sol in three. Consequently, "during our

A view of Laguna Hollow.

development period, we had this mantra that if we could get two out of three days to be successful and productive, then we'd be doing very well. With the exception of the period when Spirit had the flash memory anomaly, we've been doing much better than that." Excepting a catastrophic failure, the duration of the mission was likely to be dictated by the degradation of the power and thermal systems. As Mars approached aphelion, its heliocentric distance would increase by about 40 per cent, which would reduce the insolation. The power available would be further reduced by dust that settled on the solar panels. Also, with the onset of the southern winter, the overnight temperature would fall significantly and the thermal cycling would degrade the systems. Although the electronics were maintained in a benign environment, it was, as Cook put it, "the stuff that's outside" that was liable to fail.

Middle Ground
Spirit resumed its traverse towards Bonneville on sol 50 with a 22-metre drive, but because the rocks were becoming more plentiful and larger, this distance included manoeuvring around obstacles. The next morning it made another soil analysis for its radial survey and then drove 30 metres, setting a record for a single-sol drive. A 4-metre drive on sol 52 achieved the objective set for the previous day, which was to reach the half-way point to Bonneville, named Middle Ground, where the rover was to remain for several days. While the pancam made a 360-degree panorama to supplement that at the lander, the arm investigated the soil. On sol 53 the pancam inspected a rock named Sandia which, at 1.7 metres long and 33 centimetres high, was the largest seen to-date. Its appearance indicated that it was basaltic. Being so close to Bonneville, it was almost certainly ejecta, which did not bode well for the hope that this impact had excavated sedimentary rock. In fact, all of the sizeable rocks inspected by the mini-TES had been found to be vesicular olivine basalts. In terms of its texture, the soil was very complex; it had a crusty surface, but there were no lunar-style impact breccias – the so-called 'instant rocks' formed as the shock of an impact compresses regolith.

On sol 54 Spirit advanced 3.4 metres to a 60-centimetre-wide rock named Humphrey which possessed fractures that might have been induced by the shock of its excavation. After being examined using the microscope, it was analysed by the

Sampling the rock Humphrey: before (left), brushed (middle) and drilled (right).

APXS. The next day the RAT brushed a trio of overlapping spots in order to provide a sufficiently large area for the mini-TES to inspect, and the vehicle then withdrew to give this instrument a clear field of view. It was decided to grind just to the right of this patch, and on sol 57, having moved back into position to use its arm, the rover emplaced the APXS to confirm that the dusty surface there was the same. The 4-hour grinding operation planned for sol 58 was aborted after 20 minutes when a crack in the rock's surface caused a sensor on the RAT to indicate a loss of contact. It was decided to revise the software limits and reassign the task to sol 59, but first the APXS analysed the aborted hole because it had scraped off some of the 'rind'. A spot slightly to the left was selected for the second drilling. The material proved to be so indurated that four hours of grinding penetrated just 2.1 millimetres. The APXS analysed the hole on sol 60. Although as a fragment of basalt Humphrey was not particularly interesting, internal veins of bright material made it a significant find, because these could well be minerals that had crystallised from aqueous fluid that had seeped into crevices. "If we found this rock on Earth," noted Arvidson, "we'd say it was a volcanic rock that had a little fluid moving through it." It was decided to subject the next light-toned rock to a comprehensive investigation.

On Bonneville's rim
Spirit departed Humphrey on sol 61, and over the next several sols worked its way through the ejecta – as it approached, there was a two-fold increase in the coverage of rocks and a five-fold increase in their average size – and up the 15-degree slope onto the 4-metre-high mound of Bonneville's rim. "The terrain's getting trickier and trickier," noted rover driver Chris Leger. In manoeuvring on sol 64, the rover made an unplanned trench, which was promptly dubbed the Serendipity Trench. After examining this soil the next morning, it resumed its drive. On finally cresting the rim on sol 66, the navcam took a 180-degree look at the interior. There was no evidence of layering in the wall. On the floor, 14 metres below the rim, there was what John Grant described as "a spectacular view of drift material". Nevertheless, as mission manager Mark Adler announced, "We don't see anything sufficiently compelling to take the risk of going in there." The sampling would be restricted to the rim. On sol 67 the rover moved 15 metres around the rim to a better vista point, and while the

By sol 62 Spirit was navigating its way through the ejecta from Bonneville.

spectrometers analysed the soil there over the next several days the mast-mounted cameras assembled full panoramas. From this high point, Spirit was able to see the wreck of its heatshield beyond the far rim. On sol 71 the rover manoeuvred 18 metres anticlockwise around the rim to a strip of drift which it was to examine as a proxy for the dunes in the crater. The following day, after the mini-TES had analysed this material, Spirit advanced to put its front

On reaching the rim of Bonneville crater, Spirit returned a panoramic view of its interior.

The sinuous sandy dune on the rim of Bonneville crater named the Serpent drift.

wheels on the fringe of the drift to disturb the material, reversed to document the result, and then returned for arm work. The drift had a frangible crust that broke when crushed, and this mixed with the fines beneath. After sunset, the rover pulled away and used the mini-TES to measure the relative cooling rates, then returned at dawn to continue its *in-situ* inspection. On sol 74 Spirit drove to examine another part of the drift, and then resumed its drive around the rim.

On sol 76 Spirit reached a 2-metre block on the crest of the crater that must have been dug from the deepest excavation. The multifaceted scalloped surface of the rock indicated that it had for a very long time been scoured by sand carried by the northwesterly wind. Named Mazatzal, it was intriguing for its light-tone and sugary texture. It was decided to examine this rock in detail to determine whether it represented the putative sedimentary substrate of the basaltic plain, and on sol 77 the rover manoeuvred into position to utilise its arm. The following day it made a microscopic examination of three areas of the rock, then emplaced the Mössbauer spectrometer for an overnight integration. On sol 79 it brushed two sites, revealing a dark material under the light-toned surface, then analysed this overnight and the following day. The dark material was so hard that four hours of drilling on sol 81 penetrated only 3.8 millimetres. After the hole had been inspected, it was redrilled on sol 83 to a depth of 8 millimetres, exposing a light-toned material beneath the darker rind. After the Mössbauer had analysed the hole, a spurious signal from a contact sensor aborted arm operations, and the remainder of the sol 84 sequence had to be rescheduled for sol 85. The RAT brushed a six-spot pattern on sol 86 and the mini-TES then viewed the exposed rind to complete the study of this rock. The results showed that it had formed in the presence of aqueous fluid. "This wasn't water that sloshed on the surface. We're talking about small amounts of water, and perhaps underground," pointed out Harry McSween of the University of Tennessee. "The evidence is in the form of multiple coatings on the rock, as well as fractures that are filled with alteration material and perhaps little patches of alteration material." A bright stripe across both the dark layer and the light-toned interior was a fracture through which water flowed, precipitating minerals. By analysing the unbrushed, brushed, shallow-drilled and deep-drilled rock, the APXS provided the chemical composition of the layering. The ratio of bromine to chlorine inside the rock was unusually high, supporting the case for it having been chemically altered by water. Observing that the mini-TES showed the light-toned surface to be mineralogically different to the dark intermediate layer, McSween said that the outer layer, the dark layer and the bright veins might represent three *different* periods in which the rock was buried, altered by fluids, and exhumed, in which case Mazatzal had quite a story to tell of activity in Gusev.

It had been hoped that, as Arvidson put it, Bonneville would have "overturned the stratigraphy and exposed it to our viewing pleasure", but the fact that the ejecta was basaltic indicated that if Gusev had held a lake that deposited sediments, these must be buried more deeply than the impact had excavated. The second option was to seek a formation that was stratigraphically lower than the basaltic plain but was exposed at the surface. The Columbia Hills, some 2.3 kilometres to the southeast, were morphologically distinct from the plain. "The hills represent a different unit, likely older than the plain. There are intriguing features in the hills, and we want to investigate the processes that formed them," explained Amy Knudson of Arizona State University at Tempe. Although they might prove to be beyond Spirit's reach, much would be learned from variations in the soil and rock distribution *en route*, because there was sure to be material on the plain that derived from the hills.

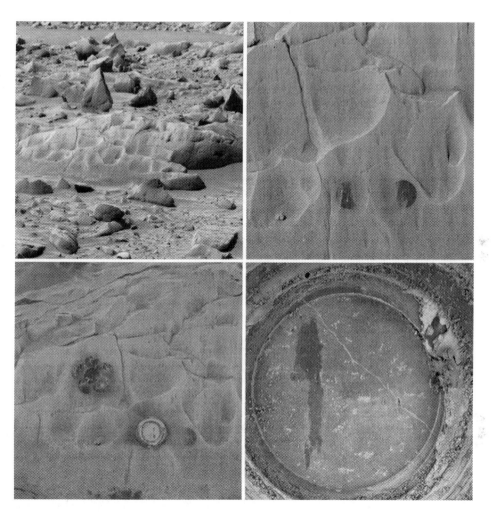

Sampling the rock Mazatzal: before (top-left), brushed (top-right), drilled and rebrushed (bottom-left) and a microscope image of the RAT hole (bottom-right).

TO THE HILLS

Spirit left Mazatzal on sol 87 and drove southeast off the rim of Bonneville to start its long trek to the hills. Although told to drive 65 metres, it had to avoid so many obstacles that by the time it was obliged to shut down it had made only half of this distance. Once it was out of the ejecta blanket, however, the pace picked up, and a routine was established in which a day began with the rover deploying its arm to inspect the soil while simultaneously using the mini-TES to inspect nearby rocks, prior to making the next leg of its drive. At the end of sol 90 (which achieved the mission-success criterion for duration) Spirit drew up beside a 60-centimetre-wide

A view by Mars Global Surveyor processed by Malin Space Science Systems to reveal the track left by Spirit as it drove to and worked on the rim of Bonneville crater.

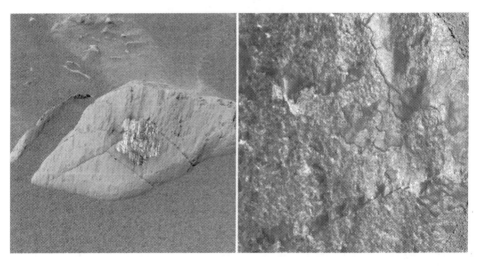

A mosaic of brushed spots on the rock Route 66 and (right) a microscope image of the cleaned surface.

24-centimetre-tall light-toned rock named Route 66. Since the vehicle was to halt for four days for its software to be upgraded, the Mössbauer was left to perform a very long integration in order to determine how this rock differed from Mazatzal. The rover was rebooted on sol 98 using the new software, which was designed to reduce the frequency with which the vehicle had to pause to snap new stereoscopic imagery to update its assessment of hazards; it was hoped that this revision would enable the automated navigation to double the overall rate of traverse to 32 metres per hour, and so greatly improve the chances of reaching the Columbia Hills.

The attraction of Route 66 was that it was shiny, with a smooth discontinuous coating that was mottled in a manner suggestive of multiple layering. A 6-spot was brushed on sol 99, and this was then analysed by the arm instruments.

Spirit withdrew on sol 101 to allow the mini-TES a clear view of the brushed part of Route 66, and then in the afternoon drove 64 metres – the longest single-sol distance to-date. It remained in place the next day while the mini-TES surveyed the rocks ahead to facilitate planning a radial approach to a crater named Missoula, then drove 75 metres on sol 103, remained in place on sol 104 to examine the rocks, and reached the rim of Missoula on sol 105. On finding the cavity to be largely filled with drift, it was decided to survey nearby rocks using the mast-mounted cameras and then resume the drive.

After several sols, Spirit spent the morning of sol 110 analysing the soil at Waffle Flats. In avoiding an obstacle on sol 112, the rover's manoeuvres resulted in it driving in reverse, using its rear-facing hazard camera to monitor its progress, and on finally halting it spun around to face forward. The next morning it excavated a trench 6 centimetres deep, and after examining this for several sols it was off again, making good progress day by day, pausing from time to time to study the soil to document any compositional changes as it approached the hills. The plan

As Spirit drove southeast from Bonneville across the plain towards the Columbia Hills, it paused at Missoula crater.

was to head for the West Spur of Husband Hill, which was the tallest of the group, spend a week characterising the transition from the plain to the hill, and then start the ascent.

After a 110-metre drive on sol 134, Spirit was within 750 metres of the base of the West Spur. A lengthy trench investigation at this point measured magnesium and sulphur in ratios indicating the presence of a magnesium sulphate salt. "The most likely hypothesis is that water percolated through the subsurface and dissolved out minerals, and as this evaporated near the surface it left concentrated salts behind," ventured Squyres. "I'm not talking about a standing body of water, but we do have an emerging story of surface water at Gusev."

As Spirit crossed the plain, it approached the West Spur of Husband Hill.

Despite having had its autonomous navigation upgraded, when Spirit resumed its traverse on sol 142 its rate of progress was restricted by the increasing coverage of rocks. It was now able to resolve features on the hill that might be outcrops, and there was a talus of rocks near the base that appeared to have rolled down. This was excellent news, because if the hill proved to be too steep for Spirit to ascend, the talus would enable it to study rocks from sites beyond its reach. "Those rocks may be the oldest material yet seen on the Martian surface," ventured James Rice of Arizona State University at Tempe. On sol 156 Spirit crossed the 'contact' from the plains unit to the hill unit, after which it started up a ramp of material that had eroded off the flank, and even before it reached the base of the West Spur the slope was already 20 degrees.

THE ASCENT

As Gusev was 14 degrees south of the equator and the Sun was north of the zenith at noon in the southern winter, the fact that its solar panel tended to face southwest on starting up the hill rendered Spirit's daily activities power-limited. In addition, its pace slowed dramatically because it now had not only to avoid rocks but also to cope with wheel-slippage on the sandy slope. A few metres up, it found a cluster of strange rocks, one of which was so intriguing that Laurence Soderblom of the US Geological Survey in Flagstaff, Arizona, was prompted to name it after "the pot of gold – the prize at the end of the rainbow". It was almost buried, had a layered appearance, and a number of small nodules were attached to the tips of narrow stalks. "This rock has a shape as if someone took a potato and stuck toothpicks in it, then put jelly beans on the toothpicks," Squyres told reporters. An examination by the microscope on sol 162 suggested that the rock had been altered by water. On sol 164, after analysing the rock's chemistry using the APXS, Spirit attempted to reposition itself to use the RAT. Initially facing south, it reversed, ran downslope on the west side of the rock, turned to face east and moved in, but ended up tilted sideways at 20 degrees and, having slipped, was 2 metres from its target. On sol 165 it slowly crept back towards the rock, trying to correct for slippage, but failed to close the gap. It tried again on sol 166, climbing while tilted, and finished up 1 metre from the rock. Another effort on sol 167 concluded with the rover straddling the rock! Only after reversing clear on sol 169 was Spirit able to apply its RAT. The tool had to knock off some nodules before it could make contact with the body of the rock, and when the force of the tool displaced the rock the drilling was curtailed. Nevertheless, the 0.2-millimetre-deep scrape was sufficient to facilitate an APXS analysis overnight and

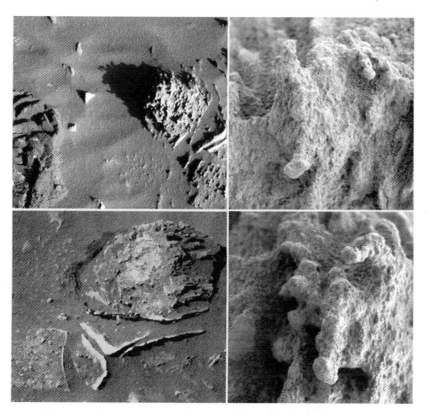

A view of the rock Pot of Gold prior to sampling (top-left), the nodules on stalks (right) and the rock after being displaced by an attempted drilling (bottom-left).

the Mössbauer the following day, which revealed the rock to contain hematite. Pot of Gold was evidently a soft material that was readily weathered except where it was protected by the nodules, and as the stalks were eroded the nodules fell off. "Hematite is pretty hard stuff," pointed out Squyres. "It's good at cementing rocks together. If you have hematite distributed through the rock in an uneven fashion, then the hematite-rich stuff might remain behind while the hematite-poor stuff gets eroded." In the case of Pot of Gold there were planar sheets of apparently resistant material running through it at different angles, and the stalks were bright and shiny where they were cut off. "There's a form of hematite – 'specular hematite' – that is very bright and sparkly," continued Squyres. 'We might be seeing that." As Pot of Gold had evidently rolled down hill, it was a fascinating teaser. A group of nearby rocks were named the Rotten Rocks because their interiors had eroded and left the more resistant rinds intact. "It's hard to imagine that water wasn't involved," noted Soderblom at an 'instant science' press conference. The working hypothesis was that such rocks formed in, or were modified by, brines. On sol 175 Spirit examined a string of shiny features that proved to be the partially exposed interiors of rocks that had fragmented as the rover ran over them. While manoeuvring to inspect this String of Pearls, the rover's wheels exposed a very bright material beneath the soil, and this was also subject to examination.

On sol 182 Spirit drove to a level site dubbed Engineering Flats that was clear of obstacles for a '3-kilometre tune-up' in which, over the next several days, it was to execute a series of manoeuvres designed to heat the lubricant in the motor of its right-front wheel, since this had recently started to draw twice the nominal current. Also, as there was a discrepancy of several centimetres in positioning the arm, this was to be 'posed' in various configurations in order to recalibrate the stereoscopic vision of the hazard camera system. After these maintenance tasks had been

Spirit in position to work on the rock Wooly Patch (directly between the wheels), the two holes that it drilled and microscopic images of the results.

accomplished, Spirit started on a northerly route up the flank of the West Spur on sol 188, but the going was tough and it achieved only 9 metres on sol 189, 13 metres on sol 190, and 10 metres on sol 191. Turning east on sol 192, it climbed 17 metres to a rocky outcrop named Wooly Patch, at which point it had run so low on power that it had to sleep to recharge its batteries. Small cracks in the surface of the rock suggested that it had been so altered by the presence of mineral-rich water that it had been softened. This was confirmed on sol 195 when the RAT drilled a 5.2-millimetre hole in only two hours. A second hole was drilled on sol 198. Gratifyingly, it was becoming apparent that the rocks on the hill were very different to the basalt on the plain.

On departing the Wooly Patch on sol 200, Spirit drove 16 metres to a position at which its solar panel would be better angled for recharging its batteries. It resumed its ascent on sol 201 heading for an outcrop named Clovis, but after several metres the slope increased to 26 degrees, exceeding the set limit of 25 degrees and prompting a halt. The next day the software was revised to raise the limit to 32 degrees, the rover set off again and completed the specified 30-metre distance. After another 19 metres on sol 203 it reached a shallow hollow whose interior slope inclined the solar panel northward for optimal recharging of its batteries. When the slippage increased to 125 per cent on sol 206 Spirit actually lost ground, and since its final position was unfavourable for illuminating the solar panel it had to remain in place the following day to recharge. In a change of plan, the rover was directed on sol 208 to make a detour around the awkward part of slope. The next day, the computer suffered an anomalous reboot as a result of a known but as yet unfixed bug in its software, and so it did not reach Clovis until sol 210. After a preliminary inspection of the weathered surface of the outcrop, a spot was brushed clean on sol 214 and drilled on sol 216. The fact that it was another soft rock was fortunate because the rover did not have the power for a lengthy grinding operation; the 8.9-millimetre hole was the deepest yet. The APXS found a much higher proportion of sulphur, bromine and chlorine than in Mazatzal, which indicated, as Squyres put it, "a more thorough, deeper alteration, suggesting much more water". Indeed, Clovis was the most highly altered rock yet found on Mars. The hypotheses included its exposure to hydrothermal fluid, volcanic gases, and a concentrated brine rich in such chemicals while underground. As the Mössbauer's gamma-ray source was decaying, this instrument had to be left in the hole for all of 48 hours to build up a high signal-to-noise ratio. The Mössbauer identified goethite – a mineral that has hydroxyl in its crystalline structure and was strong evidence for past water. Sulphur, chlorine, bromine and potassium are readily transported by water. Their presence in abundance indicated that the rocks on the hill had once been infused with water. As a final act, the RAT brushed a 7-spot mosaic and Spirit reversed to enable the mini-TES to inspect the result.

On sol 226 Spirit drove 8 metres to a rock named Ebenezer, which it brushed on sol 230, drilled to a depth of 3.4 millimetres on sol 231, and brushed to make a 7-spot mosaic for the mini-TES on sol 236. By now, it was clear that the rock exposures on the hill were old and weathered. "We haven't found a single unaltered volcanic rock since crossing the contact from the plain to the hills," observed Squyres, "and I'm beginning to suspect we never will."

A mosaic of brushed spots and a hole in the rock Clovis, and a microscope image of the hole (right).

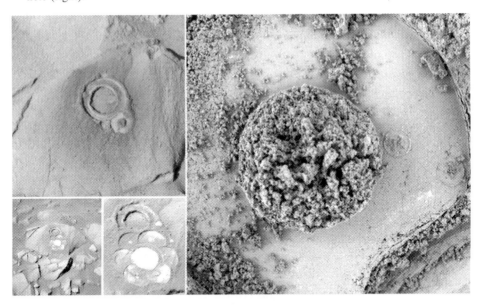

Spirit works on the rock Ebenezer.

On sol 238 Spirit drove 9 metres to a convenient location at which to spend the period in September when communications were disrupted by the passage of Mars behind the Sun as viewed from Earth. On resuming work on sol 263, it drove to a point on the southern rim of a 2-metre-diameter depression, and the following day slowly manoeuvred around this slope, matching the progress of the Sun across the sky, to recharge its battery. The drive on sol 265 was aborted by a signal from one of the wheels, and while this was investigated the arm inspected nearby soil. On concluding that the signal had been spurious, the traverse was resumed, and on sol 271 Spirit drew up in front of Tetl – the first horizontally layered rock to be seen at Gusev – which proved to be volcanic material altered by abundant water. On sol 277 Spirit set off to another layered rock 6 metres away named Uchben, but was halted half way by another warning. The signal was related to the right-front and left-rear wheels, which were two of the four steering wheels. To avoid losing more valuable time, the rover resumed its drive on sol 281 with the two 'faulty' wheels unpowered, and finally reached Uchben. A diagnosis determined that although the signal indicated that the brakes had failed to disengage, in fact they had done so and the signal to signify this was impeded by an accumulation of insulating material on the contacts designed to report disengagement. As a 'workaround', the software was revised to ignore this spurious signal. The RAT drilled a 6-millimetre hole in Uchben on sol 285, and when the microscope inspected the result it revealed sand-sized particles. The angularity of some of the particles implied deposition as an ash fall, but there were others whose rounded shapes suggested they had been transported across the surface by wind or water. "We've really made headway in the last few weeks in understanding these rocks," noted Squyres. Next, Spirit set off to examine nearby rocks, including one, named Lutefisk, which had nodules on it. By now, the sticky front wheel had ceased to draw excessive current, demonstrating that the maintenance at Engineering Flats to redistribute its lubricant had been successful. On cresting the ridge named Machu Picchu at an elevation of 40 metres on sol 325, Spirit got its first view of its route across to the flank of Husband Hill.

On sol 332 Spirit examined a rock named Wishstone whose surface possessed a 'jumbled' texture. It proved to be very rich in phosphorus in the form of apatite – calcium phosphate. The basaltic fragments randomly distributed in the light-toned matrix suggested that it was the product of a violent volcanic explosion. A nearby rock named Wishing Well was similar. Squyres told reporters that these were "a completely new type of rock, unlike anything seen before on Mars".

A depiction of Spirit's route as it ascended the West Spur of Husband Hill.

Microscopic image mosaics spanning the vertical extent of the layered rock Tetl.

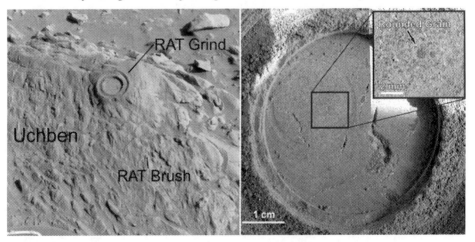

Spirit works on the rock Uchben.

The flank of Husband Hill was very sandy and littered with small pebbles, with the result that although the slope was less than 20 degrees the rover began to suffer very high slippage, and on sol 339 it had to halt when a potato-sized rock became lodged between the inner part of the right-rear wheel and its drive mechanism. The rock was shaken free on sol 340, but in doing so the wheel dug itself into the sand,

Spirit works on the rock Wishstone.

and whilst the rock was on the ground it was still inside the rim, where it remained until the vehicle managed to extricate itself on sol 346. On sol 350, Spirit was told to drive 5 metres upslope to a rock named La Brea, but slippage prevented it from achieving more than about 1 metre in the time available – on the hillside its solar panel was not favourably illuminated. After fairing little better the next day, it was told to examine a rock named Champagne that fortuitously was within reach of the arm, and this was drilled on sol 355. On sol 358 the rover attempted to resume its ascent to the next ridge. After several days of slippage, sometimes exceeding 100 per cent, it finally reached firmer ground and gained 45 metres in four days, then on sol 371 reached an exposure of bedrock named Peace. The drilling on sol 374 was restricted to 40 minutes because of power limitations imposed by a recent dust storm, but the tool was able to penetrate the rind to a depth of 3.2 millimetres, and several days later this hole was deepened to 9.7 millimetres. The analysis showed Peace to contain a higher proportion of sulphate salt than any rock yet examined. "Usually, when we have seen high levels of sulphur in rocks at Gusev," explained Ralf Gellert of the Max Planck Institute for Chemistry in Mainz, leading the APXS team, "it's been at the very surface of a rock. In this case, the sulphate is deep inside." Other data showed the rock also contained significant amounts of olivine, pyroxene and magnetite. "This is probably the most interesting and important rock that Spirit has examined," Squyres informed reporters. "It may be what the 'bones' of the hill are made of. It's even more compelling evidence for water playing a major role in altering the rocks here." Discussing the texture of the rock, he said it was "as if

Spirit works on the rock Peace.

volcanic rocks which had been ground into little grains were cemented together by a substantial quantity of magnesium sulphate salt". The salt could have come from percolating liquid water containing dissolved magnesium sulphate that slowly evaporated and left the salt behind, or it could have come from magnesium-rich minerals already in the rock being weathered by an infusion of dilute sulphuric acid.

Spirit resumed its ascent of Husband Hill on sol 381, with a 21-metre drive up the 18-degree slope that left it in position to investigate a rock named Alligator. It set off again on sol 388, and at the end of sol 394 parked conveniently oriented for aiming its high-gain antenna for an upgrade in its flight software. When the rover's wheels stirred up very bright soil on sol 398, it was decided to pause to examine it. This was, as Squyres put it, "an absolutely serendipitous discovery". The material, which was named Paso Robles, had the highest salt concentration of any rock or soil yet examined anywhere on the planet – it was "more than half salt", mainly a hydrated iron sulphate, but there was also a high fraction of phosphorus. "Clearly," Squyres said delightedly, "with this much salt around, water had a hand here." On sol 407 Spirit reached Larry's Lookout on the crest of the Cumberland Ridge, from where, over the next several days, it returned a panorama that gave the first view of the Tennessee Valley beyond, nestled between the hills. With the atmosphere finally clearing of dust, the power situation began to improve and, remarkably, on sol 420 a

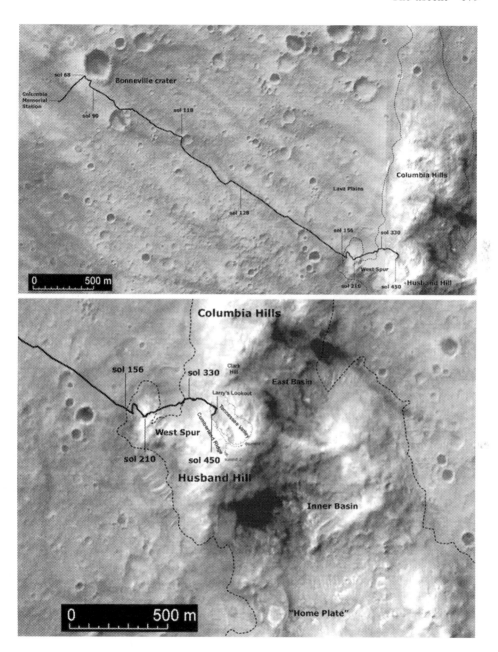

Spirit's route across the Gusev plain and ascent of Husband Hill.

Faced with this litter of rocks, Spirit had to abandon its ascent of Husband Hill.

dust devil seemingly passed over the vehicle and restored most of the 40 per cent reduction in the noon power output due to the progressive accumulation of dust on the solar panel. This significantly improved the prospect of Spirit continuing to the summit of the hill. The following day, running its hazard cameras in 'movie mode', Spirit caught two dust devils about 1 kilometre away, travelling across the plain at a speed of about 3 metres per second. Continuing its ascent of Husband Hill, Spirit entered an area in which, although the slope was only 12 degrees, there was a litter of sizeable rocks that were 'floating' on loose sand, and had to employ the tactic of 'switchbacking' to climb further. When Spirit was denied its main relay link by a fault on Mars Odyssey on sol 448 it had to remain in place, but the satellite service was restored four days later. Unfortunately, the next day the rover suffered a spurious reboot and remained out of action until sol 452.

Meanwhile, it had been decided to abandon the slope ahead, and Spirit was told to run down cross-slope until clear of the sandy terrain and renew its ascent using a better route, but when an overhanging outcrop named Methuselah was observed, Spirit was diverted, and on arrival on sol 467 it proceeded to utilise its full panoply of sensors to investigate a section named Keystone. A 4,000 by 6,000 pixel mosaic by the microscope – the most frames ever taken of a single feature – showed Keystone to be finely layered. Since its texture was similar Peace, the rock was expected to be rich in sulphates, but it proved to resemble Wishstone in being rich in phosphorus. On noting that the strike and dip of the layering of this outcrop closely matched the slope of the hill, it was realised that the flank of Husband Hill was a single stratigraphic unit. "That was the critical moment, when it all began to fall into place," mused Squyres. "Looking back downhill, you can see the layering, and it suddenly starts to makes sense." It was decided to postpone further ascent of the hill and instead examine other parts of the outcrop. Although rocks in different layers shared compositional traits which suggested a shared origin, their stratification and textures varied, as did the degree to which the minerals had been chemically altered by exposure to water or other processes. "Our best hypothesis is we're looking at a stack of ash or debris that was explosively erupted from volcanoes and settled down in different ways," Squyres reported following several weeks of intensive investigation. "Once upon a time, Gusev was a violent place – big explosive events were occurring, and there was a lot of water around."

Upon manoeuvring to find a more viable route up Husband Hill, Spirit spied this overhanging outcrop named Methuselah.

By sol 500, Spirit resumed its drive around the flank of Husband Hill, in search of a passable route up to its summit.

REFLECTIONS ON GUSEV

Spirit investigated the substructure of the Gusev plain by sampling the rock ejected by the impact that created Bonneville crater. It then drove 2 kilometres across the plain to the Columbia Hills and analysed the variations in the composition of the rocks. In this way, it provided a glimpse of the 'big picture' that would otherwise have eluded a static lander. The morphology of Gusev strongly suggests that it once held a substantial lake – after all, a long sinuous channel runs down from the southern highlands and breaches its southern wall, and the opposite rim is etched where water spilled over and drained further down the slope towards the line of dichotomy. However, instead of finding the floor of the crater to be a sedimentary carbonate or evaporite, Spirit found only volcanics, imply-ing that the plain was an ancient lava flow.

This Mars Global Surveyor image shows Apollinaris Patera north of Gusev crater (with Spirit's landing site marked at the bottom edge of the frame).

Although some of the rocks had been chemically altered by water, this did not require them to have been immersed; the aqueous fluid could have been present as volatiles in the lava, or it could have percolated through the rocks while underground. The rocks on the plain therefore provided no evidence to support the lake hypothesis. Fortunately, Spirit was able to reach the Columbia Hills. The fact that the peaks rise 100 metres above the plain did not rule against their being of sedimentary origin, since Gusev's floor lies 2,500 metres below its rim. Although Spirit's ascent of the West Spur and Husband Hill revealed the material to be of volcanogenic origin, most likely as either an airborne ash fall or a pyroclastic flow from Apollinaris Patera 250 kilometres to the north, the degree to which the rocks had been altered by aqueous fluid held out the prospect that the hills had once been submerged in a lake. As for the dilemma of the absence of carbonate, as Spirit was exploring Gusev this mystery was being resolved on the opposite side of the planet by its twin.

8

Opportunity

MERIDIANI

Meridiani Planum overlies the ancient cratered terrain of Arabia Terra. A study by Brian Hynek, Raymond Arvidson and Roger Phillips of Washington University in St Louis interpreted it as a blanket of pyroclastic several hundred metres thick. The appearance of the craters on the plain suggested that they are either being rapidly eroded or have been buried and are in the process of being exhumed. In 1998 the thermal emission spectrometer on Mars Global Surveyor identified the signature of grey hematite on a particularly flat section of this high plain.

"There are several hypotheses to explain this hematite," said John Grant of the National Air and Space Museum in Washington, DC, in a press briefing. It could have been produced directly from iron-rich lava, a process that would not require liquid water. If a mineral called goethite were to be found alongside the hematite, it would strongly imply that the hematite formed in wet conditions. On the other hand, if magnetite was found and goethite was not then the involvement of water was unlikely. If water was involved, the leading theories were iron-rich water in a lake and hydrothermal fluid percolating through volcanic ash. Simply being able to observe how the hematite was distributed would provide some insight: that is, if it was embedded in a stack of layers it probably formed on the bed of a lake, but if it was present as veins in rocks it was more likely to have formed in groundwater. This was to be investigated by Opportunity, the second Mars Exploration Rover.

TWO FOR TWO!

As Opportunity approached Mars on Saturday, 24 January 2004, Peter Theisinger, the project manager, warned that although Spirit had demonstrated that the entry, descent and landing system *could* place a package on the surface, the system was complex and it was possible that the second would fail. The entry occurred in darkness over Valles Marineris. Although the low-level wind deflected the lander northward, the logic of the DIMES system decided not to fire the transverse rockets, and the main retrorockets brought the lander to a complete halt prior to severing the

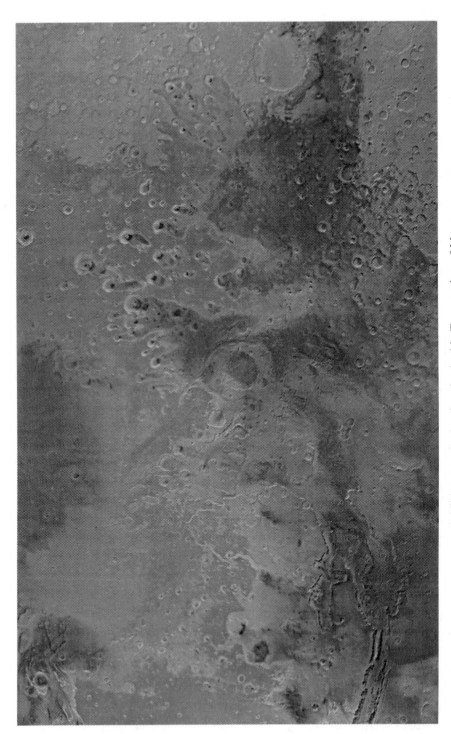

A Viking mosaic showing the Arabia Terra region of Mars.

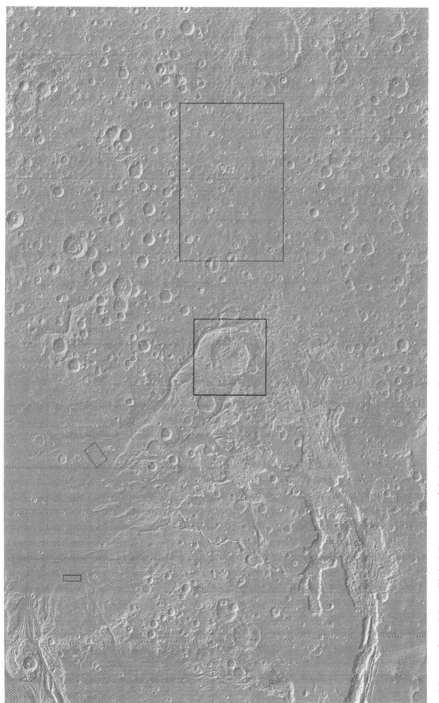

Altimetry from Mars Global Surveyor of the Arabia Terra region, with (left to right) boxes for Viking 1, Mars Pathfinder, Aram crater and Meridiani Planum.

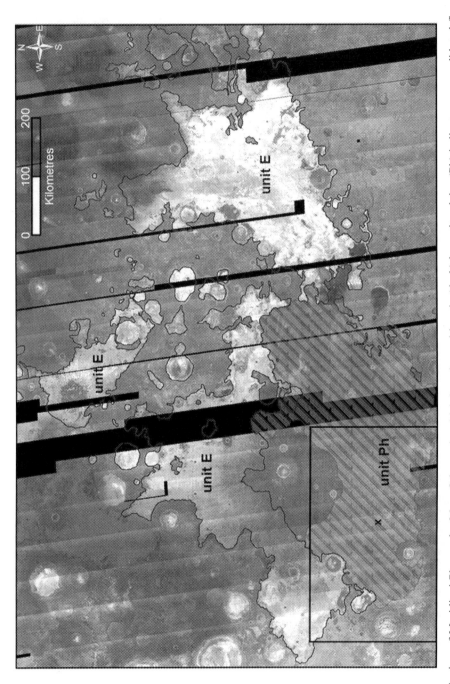

A view of Meridiani Planum by Mars Odyssey in which low thermal inertia (dark) hematite plains (Ph) indicates unconsolidated fine grains and high thermal inertia (bright) 'etched' (E) terrain indicates mixed surfaces with increasing induration and rock abundance (adapted from B.M. Hynek, *Nature*, vol. 431, 2004).

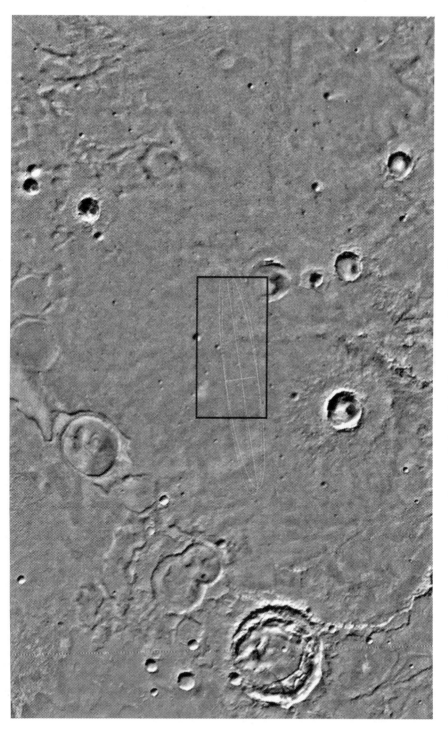

A general view of the Opportunity landing ellipse.

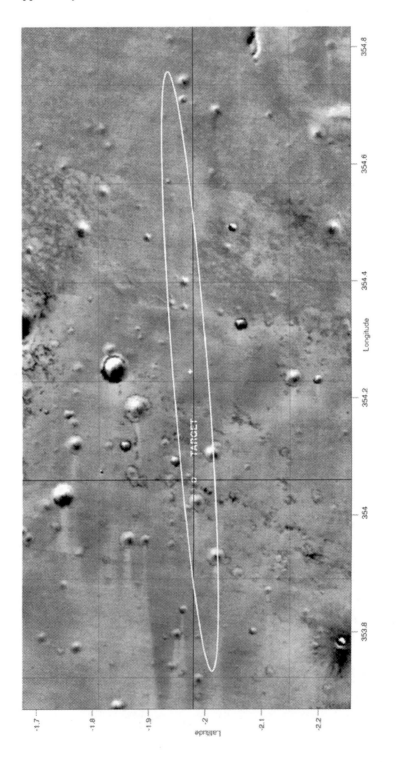

A close-up view of the Opportunity landing ellipse.

bridle, with the result that the 2.5-g first contact was well within the 40 g shock the system was designed for. The intermittent direct signal created the impression that the lander rolled for 20 minutes, but when Mars Global Surveyor relayed recorded telemetry it became apparent that the lander had come to rest within a few hundred metres, with its antenna facing down, and reflections from the ground had caused the strength of the signal to vary. After deflating its airbags, the lander used the leverage of its side petals to right itself, and the rover deployed its appendages and made its initial observations.

A 'hole in one'
When Mars Odyssey made the first UHF relay pass four hours after Opportunity's landing, it received 20 megabits of data that included 77 images. "Opportunity has landed in a bizarre, alien landscape," said lead scientist Steven Squyres of Cornell University as the navcam frames were presented at the following press conference. The site was virtually featureless apart from a horizontal outcrop of light-toned slabby rocks, but this was astonishing. "Scientifically speaking, this is the sweetest spot I have ever seen!" The proximity of the horizon prompted Squyres to venture that the lander had rolled into a small crater. The soil seemed to have a consistency of talc, and was so fine that it had perfectly preserved the imprints of the airbags.

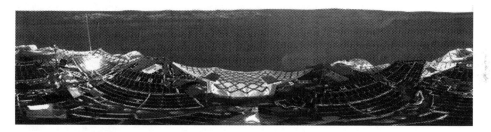

Opportunity's first navcam sequence revealed an astonishing landscape.

Having recharged its batteries, Opportunity went to sleep at sunset, but awoke during the night for a second relay via Mars Odyssey, and downloaded the DIMES imagery and the first colour pictures from the pancam. In colour, the soil appeared to be a mix of very finely grained dark-reddish material and coarse dark spherules whose colour led to speculation that it was these that contained the grey hematite. In view of the fact that the individual crystals in a grain of grey hematite are red, one idea to account for the absence of spherules in the imprints of the airbags was that they were pulverised. Another possibility was that they were sufficiently solid to have simply been driven down into the soil by the weight of the lander.

"We knew there were two fundamental geological units here," Squyres said at the next press conference. "One is a thick sequence of layered rock, fairly light in tone. We don't know what the layers are. And then on top of this is a thin coating that contains the hematite. My fondest hope after looking at the pictures from orbit before we landed, was that we'd land some place that would be close enough to a crater that we'd have a chance of traversing to it and actually getting to the layered material. Instead, what's happened is, we scored a 300-million-mile interplanetary hole in one and landed in a small crater!" The crater was 22 metres in diameter and 2 metres deep, but the rover was not expected to encounter any problem in driving out onto the surrounding plain. In view of the fact that on its interplanetary cruise Opportunity had made only three mid-course corrections, and since a 'hole in one' for a 'par 3' in the game of golf is known as an 'Eagle', the crater was so-named. Opportunity had hit the jackpot, because the outcrop appeared to be an exposure of *bedrock* and, as Allan Treiman of the Lunar and Planetary Institute in Houston told journalists, "rock that hasn't moved since its formation is supremely important to a geologist because it helps to explain the large-scale structures and processes". Given the odds against landing in a crater having an outcrop of bedrock in its wall, this feature was informally dubbed the Dumb Luck Formation. "To understand the chemistry and geological history," added Darby Dyar of Mount Holyoke College in Massachusetts, "we need to study rocks showing the fewest signs of weathering. Any loose sediment or boulder has – by definition – been released from bedrock as a result of either chemical or physical weathering processes and so is likely to have been altered since its formation. Access to bedrock gives us a chance at analysing *unaltered* rock." The excitement was palpable. The outcrop was 8 metres from the lander, and ranged between 30 and 45 centimetres in vertical extent. Most of it was on a shallow slope, but some slabs had near-vertical faces. As the pancam worked its way along, it resolved a series of thin layers, some as fine as 1 centimetre thick. "These aren't lava flows," Squyres told reporters "they're something we haven't seen before on Mars." They appeared to be sedimentary, but it was not yet possible to state whether the sediment had been emplaced by water, by wind, or as a fall of volcanic ash. The rover's microscope would be the key to resolving the issue. Ash particles would be highly angular. If the material was deposited by either the wind or surface water then the particles ought to be rounded. Andrew Knoll of Harvard pointed out that, to a geologist, layered rocks were like the pages in a history book – they told the story of how the rock was formed and subsequently altered.

Meanwhile, the entry, descent and landing team analysed the DIMES imagery in an effort to precisely identify the landing site. In the case of Spirit, this had been straightforward because it was possible to take bearings on hills, but Opportunity could barely peer over the rim of the crater in which it had come to rest. A number of candidates were suggested, all downrange, up to 24 kilometres from the middle of the 75-kilometre-long target ellipse. It proved possible to locate Eagle crater by analysing the UHF signal received by the orbiters, at which time it was noted that there was a much larger crater nearby that would have excavated more deeply into the bedrock. "The way I envisage this mission going", Squyres announced, "is we drive off the lander and look at the soil to investigate the hematite mystery; we go to the outcrop because it's right there in front of us, and explore it in some detail to understand that geologic unit; then we climb out of the crater, take a look around, and head for the big crater. It's going to be a wonderful adventure."

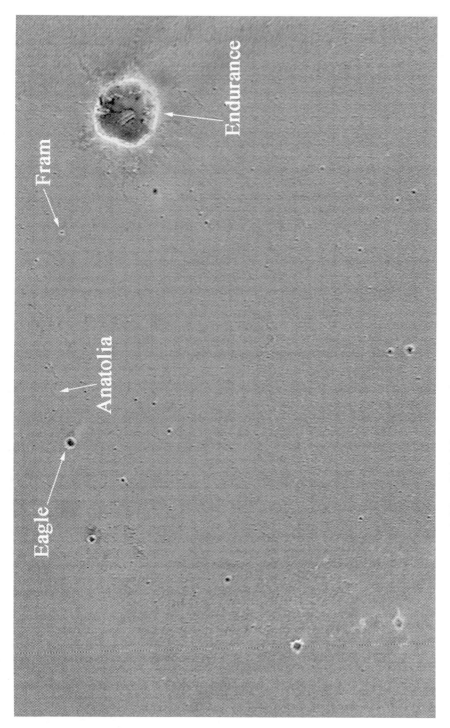

A Mars Global Surveyor view of the Opportunity landing site.

At this point in the mission, when it was 13:00 at JPL on Earth, it was 17:00 at Gusev and 05:00 at Meridiani on Mars. Consequently, operations would have to be conducted around the clock. Two operations teams were formed, red for Spirit and blue for Opportunity, sharing the Surface Mission Support Area in 12-hour shifts. It would be exhausting, since 'off duty' scientists would inevitably attempt to keep up with the progress of their counterparts.

On the anniversary of the loss of the STS-51L crew on 28 January 1986, the landing site was named the Challenger Memorial Station.

First impression

The base petal was fairly level, being just five degrees off horizontal, and the rover was facing north-northeast. On Thursday, 29 January, the airbags at the rear of the lander were further retracted and that petal angled down onto the surface to tilt the base and dip the exit ramp in order to give Opportunity what mission manager Matt Wallace described as "an easy ride" off the lander. On Saturday, as a reward for completing the 'stand up' two days ahead of schedule, Kevin Burke, the chief engineer of the post-landing-to-egress phase, was given the privilege of issuing the command to drive off the lander. A few hours later, Mars Odyssey relayed a view from the rear hazard camera that confirmed it to be on the surface. As had Spirit, Opportunity was to remain at the egress point for several days methodically testing its instruments.

The first mini-TES analysis was presented at the post-egress press conference, and it confirmed that the hematite was associated with the spherules. "One of my worst fears", reflected team leader Phil Christensen of Arizona State University in Tempe, "was that the hematite was in the dust and was in *everything*, as then we'd have a hard time tracking down the source. It was extremely gratifying to see that it's in this coarsely grained material. I believe the source was the rock unit which once sat on top of the outcrop." It was evident from the narrow arcs that the mast-mounted cameras were able to see over the rim of the crater, that if the light-toned bedrock extended far and wide then it was exceptionally flat and thinly mantled by soil and hematite spherules. It was estimated that several metres of rock must have been eroded off to leave such a 'lag' of spherules. When Opportunity examined the soil, it proved to be rich in olivine. On the basis of orbital imagery showing light-toned rims of the craters on the dark plain, Christensen argued that the outcrop was very thin, and that the impact that made Eagle crater had punched right through to a basaltic substrate, the erosion products of which were filling the hole. Of course, if the light-toned rock was considerably thicker than the depth to which this impact had excavated, the olivine-rich sand in the crater must have been blown in by the wind.

SAMPLING THE OUTCROP

On sol 12 Opportunity drove in a series of arcs, two to the left and one to the right, then made a 30-degree turn-in-place and advanced 1.8 metres to position itself 3 metres nearer the right-hand end of the outcrop. These manoeuvres served to assess its mobility on the unusual surface. On sol 13 it moved 1.1 metres closer towards a knob of rock (dubbed the Snout) on the outcrop, but its wheels slipped in the loose material as the slope increased to 13 degrees, and it lost ground. It drove another 35 centimetres upslope the next day into a position from which it could inspect the Snout (later renamed Stone Mountain).* The first task was an inspection of the soil at the base of the outcrop for comparison with that at the egress site – the fact that it had more sulphur suggested that the rock above was a salt-containing sediment. When the microscope inspected the rock, it was revealed to be a stack of layers, each of which was only a few millimetres thick. "Embedded in it, like blueberries in a muffin, are little grey spherules," said Squyres, thereby coining the nickname by which the spherules would popularly become known. The spherules were seen in various stages of eroding from the fine-grained matrix, presumably as a result of wind action. When the composition of the outcrop was analysed by the Mössbauer and APXS the rock was confirmed to be rich in sulphur.

The Stone Mountain section of the outcrop.

The plan was to traverse along the base of the outcrop – now named Opportunity Ledge – pausing about every 6 metres to enable the pancam and mini-TES to document it at the highest resolution, in preparation for selecting targets for later examination. However, because the slope comprised loose sand and spherules the wheels suffered significant slippage and on the first leg, on sol 16, it managed only 4 metres. The next day it examined the rock at this site, dubbed Point Bravo. When the robotic arm failed to stow properly, the software cancelled the drive to Point Charlie that had been assigned to sol 18, but this was done on sol 19. This initial survey of the outcrop revealed that the layering tended to converge and diverge at shallow angles, suggesting that the rock was formed in the presence of a moving current. Of course, there were several hypotheses, including volcanic flow, wind and water. On sol 20 Opportunity was to investigate the soil at Point Charlie, but the software intervened during the emplacement of the microscope: to continue would have violated a constraint on arm motion – the action had been tested on the engineering vehicle at JPL, but the replica of the situation had not been sufficiently accurate. It had been intended to follow up with a 9-metre drive to Hematite Slope, which the mini-TES had shown to be particularly rich in hematite, but the aborted

* See Colour Plate 6 for an annotated view of the outcrop.

action left the arm unstowed, which precluded driving. As the command sequence for each sol was uploaded when the rover awoke soon after sunrise, the arm could not be recovered until sol 21, but once this had been done the drive was made. The rover spent sol 22 documenting the site that had been chosen for trenching, and the next day spun its right-front wheel alternatively forwards and backwards, making a slight turn between sessions, in order to scrape an excavation 50 centimetres long, 20 centimetres wide and 10 centimetres deep. It was "a patient, gentle approach to digging", said Jeffrey Biesiadecki, who planned the task, with 6 minutes of wheel action over a 22-minute period leaving pauses to allow the motor to cool down.

On sol 23 Opportunity used its wheels to excavate a trench.

"It came out deeper than I expected," admitted Robert Sullivan of Cornell. It was immediately noted that the soil in the upper part of the wall of the trench had a 'clotty' texture, and that the soil on the floor was brighter than the dark material on the surface. Sullivan speculated that the texture of the surficial layer might indicate the presence of adhesive salts. On sol 24 the arm was deployed to determine the mineralogy, chemistry, and texture of the various materials. "What's underneath is different to what is at the immediate surface," noted Albert Yen of the JPL science team. Although there were small shiny pebbles beneath the surface, the soil was so finely grained that the microscope could not resolve its constituent particles. While in place performing this arm work, the rover was also downloading the backlog of imagery that it had taken during its traverse of the outcrop. At this point in the MER mission, the two rovers, between them, had returned 10 gigabits of data: 66 per cent was relayed by Mars Odyssey, 16 per cent by Mars Global Surveyor, and the remainder was sent directly via the high-gain antennas to Earth, but the X-Band transmitter had to be used sparingly since it drew more power than the UHF relay.

On sol 26 Opportunity left its trench and drove back along the outcrop, halting 2 metres from a section named El Capitan, which was documented in detail by the pancam and mini-TES the following morning. "The whole stack appears to be well exposed here," Squyres pointed out, which was why this section had been selected. In the afternoon the rover eased 30 centimetres up the slope to a position at which it could use its arm. Over the next two days, a large number of microscope images were taken to be mosaicked to record the structure in exquisite detail. The spectrometers were operated in turn overnight. Two grinding sites were selected: Guadalupe near the top of the stack, and McKittrick lower down. On sol 30 Opportunity deployed its

The El Capitan section of the outcrop, showing McKittrick (left) prior to and after drilling; and (top) Guadalupe, the drilling target, the hole and a microscope image showing blueberries within the rock (lower right).

RAT for the first time. Running intermittently over a period of two hours in order not to overheat, the tool drilled into McKittrick to a depth of 4.3 millimetres, then stalled on dislodging a blueberry that was embedded in the rock. Although this hole was not as deep as planned, it was found on microscopic inspection that the tool had cut into two other berries and exposed their interiors, which was a welcome bonus. The APXS was inserted into the hole the next day, followed by the Mössbauer, which integrated for 24 hours to secure data for a high-resolution mineralogical analysis. On sol 33 the rover eased 15 centimetres further upslope for better access to Guadalupe, and the next day this was drilled to a depth of 5 millimetres. The rover reversed on sol 36 to enable the mast-mounted cameras to document the two work sites. "We have high expectations of understanding the extent to which the outcrop has been modified chemically, and whether water was involved," Arvidson told reporters, adding that there were two working hypotheses for how the layered rock was formed. "One idea is that it's an ash fall – or simply windblown material – that became compacted; another is that it's associated with sedimentation in an old lake or shallow sea."

Opportunity drove 4.25 metres along the base of the outcrop towards the Big Bend section on sol 37, but slipped on the slope, which was 22 degrees in places, and halted 30 centimetres short of its target. The next morning, it deployed the arm to sample the soil at that point, then closed in on a rock named Last Chance, which it examined over the next two days.

Meanwhile, the El Capitan results had been analysed, and at a press conference on 3 March it was reported that the rock contained sulphates that *must* have been created by a process involving liquid water, and that, as Squyres put it, the site had once been "drenched" with water. While this was the 'smoking gun', it was not yet possible to say whether this water stood on the surface or percolated underground. There were multiple lines of evidence. Firstly, there was a very high concentration of sulphur at the McKittrick site. "The chemical forms of this sulphur appear to be in magnesium, iron or other sulphate salts," explained Benton Clark, chief scientist for space exploration at Lockheed Martin in Denver, Colorado, which had built the rover. In fact, the sulphate salts made up 50 per cent of the rock. In addition, there were elements that can produce chloride and bromide salts. The clincher, however, was the presence of jarosite, a potassium–iron sulphate whose crystalline structure includes hydroxyl, whose creation *requires* the presence of water since it forms by the chemical alteration of basalt by *acidic* aqueous sulphate. "There's much more to it than just saying 'water was there'," Squyres explained to reporters. "We have made some significant steps towards *characterising* the water environment." The water was acidic because it had been infused by the emanations from volcanoes. As regards the blueberries, the microscopic imagery of Guadalupe had shown these to be distributed throughout the rock, rather than confined to distinct layers as would have been the case if they had originated elsewhere, been deposited on the surface, and then buried by later sedimentation. The fact that the berries had not deformed the layering of the rock indicated that they had not been impressed into soft sediment. This ruled out the suggestion that they were lapilli – tiny droplets of molten volcanic or ejecta glass. "Had the spherules been tossed in by a catastrophic

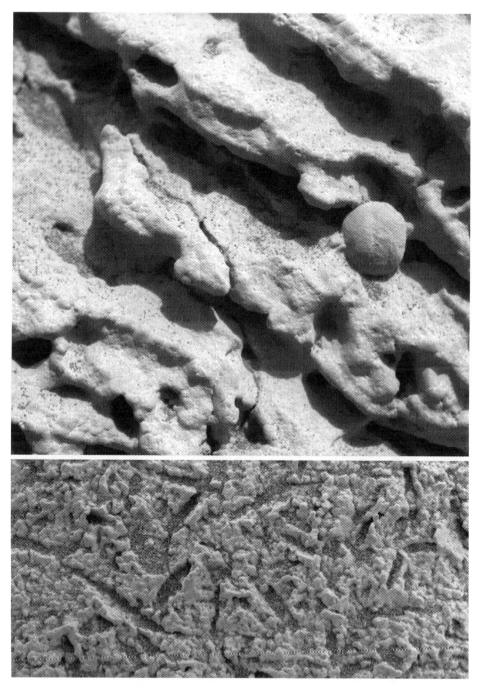

Microscope images showing (top) the fine layering of El Capitan with a blueberry in the process of eroding from the rock, and (bottom) the slot-like voids.

event, they'd not be distributed so randomly," said John Grotzinger, a sedimentary geologist at the Massachusetts Institute of Technology. The fact that linearities in the surrounding rock continued through the blueberries showed that they had formed *in situ*. Water that was rich in iron-bearing minerals had evidently percolated through the rock, and the minerals had precipitated out of solution and accreted around an irregularity, such as a grain of sand, in the sulphate. Similar spherules are found on Earth: some in Utah formed after groundwater that had percolated through iron-rich material later precipitated the iron in sandstone. The final line of evidence was that El Capitan was riddled by indentations shaped like disks about 1 centimetre in length and a fraction of this in width, positioned in random orientations. On Earth, such 'voids' indicate where crystals that formed in rock exposed to briny water were subsequently either dissolved by less salty water or eroded away. In this case, the geometry of the voids was suggestive of evaporite minerals.

"NASA launched the MER mission specifically to check whether at least one part of Mars ever had a persistently wet environment that could possibly have been hospitable to life," pointed out James Garvin, who was the agency's chief scientist for Martian exploration. "Today we have strong evidence for an exciting answer – yes!"

Opportunity returned to the centre of the outcrop on sol 40, and the following morning, after studying a section named The Dells, moved to Slick Rock near the far end of the exposure. As the RAT tried to grind a hole the next day it failed to make contact with the rock. After diagnostic tests verified the tool to be functional, the task was reassigned to sol 44, but the motor stalled after drilling to a depth of 3.1 millimetres; nevertheless, the composition was able to be analysed overnight. On sol 45 the rover repositioned itself to inspect a cluster of blueberries that had rolled into a shallow depression on the upper surface of a nearby rock. The mini-TES had shown the hematite to be present in the berries, but because the 'windows' on the arm's spectrometers were larger than an individual berry it had not been feasible to analyse their composition in detail. An analysis of the berries in this Berry Bowl confirmed that hematite was the primary iron-bearing mineral.

Having concluded its examination of the outcrop, on sol 51 Opportunity began a five-stop survey of the soil across the centre of the crater and up the opposite wall, in order to characterise the variations in sizes and shapes of the constituent particles. "We're seeing the effects of differences in wind speed," said Bethany Ehlmann of the University of Washington in St Louis. "In some patches more than others, the wind has removed small particles and left behind the larger ones." Late on sol 56 the rover attempted to drive over the rim of the crater, but its wheels slipped and it fell short. The next morning it turned right and drove cross-slope, crested the rim, and advanced 9 metres across the plain beyond, where it halted and took a navcam panorama that finally showed its owners the remarkable singularity of this landing site.

Opportunity remained in place for several days to examine a light-toned patch of soil, named Mont Blanc, that was prominent on the darker soil and proved to be the same type of dust as that found at all previous sites – it had been blown in by the wind. Meanwhile, the other mast-mounted cameras took panoramic sequences.

The Berry Bowl, where a number of blueberries had collected in a concavity on a rock in the outcrop.

Because each pancam frame was shot through six filters, the resulting 558 images used 75 megabytes of flash memory, and the matching mini-TES sequence could not be undertaken until some memory was freed. When mosaicked, this Lion King Panorama explained why the bedrock was exposed on only one side of the crater – the prevailing wind had swept that rim clean, and built up sand on the opposite side.

Meanwhile, the science team had been analysing the microscope imagery of the light-toned sulphate-rich rock in order to determine whether this had been formed on the surface. They were seeking 'crossbedding', where layers vary in thickness along their length as a result of erosion during deposition. Although crossbedding can result from either wind or water action, these can be distinguished by tell-tale structures in the layering. An exposure of rock that was formed from material laid down in water that was flowing on the surface will tend to show the sinuous crest lines of underwater ridges.

"We think we're parked on what was once the shoreline of a salty sea," said Squyres on 23 March, launching the press conference to report the analysis of the very fine layering in the Last Chance part of the outcrop. The study was based on a stereoscopic analysis of mosaics of images taken by the microscope. In addition to crossbedding, there were curves known as festoons that are typically created when a water current sifts loose sediments. "Some patterns might have resulted from wind action, but others are clear evidence of flowing water," explained Grotzinger. In this

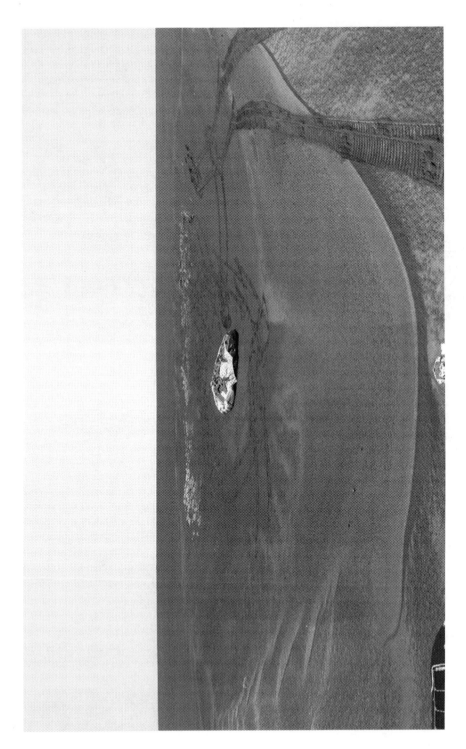

Having exited Eagle crater on its second attempt, Opportunity took a final look at its lander before setting off to explore.

case, the shapes indicated that the water had been at least 5 centimetres deep, and perhaps much deeper, and had flowed at the gentle pace of 10 to 50 centimetres per second. A second line of evidence was the presence of chlorine and bromine, which indicated that the rock was once immersed in a salty fluid that evaporated. This implied a shallow sea that sometimes dried up, leaving behind a salt flat, or playa. The earlier finding that the rocks had been drenched in mineral-rich water had not resolved whether the water was present as the rocks formed, or had altered them after their formation, possibly underground, but this new evidence indicated that they had formed *on the surface*. The bromine, in particular, strengthened the case for the rock-forming grains having precipitated from fluid on the surface as the water evaporated and the salt concentration exceeded the saturation point. "I think the explanation that the team put together is the best explanation for those rocks,"

The Last Chance section of the outcrop, showing a form of crossbedding known as festoons that are typically created when a water current sifts loose sediments.

agreed David Rubin of the US Geological Survey at Santa Cruz in California – a sedimentologist who had been asked to provide an independent assessment: "The sedimentary structures were deposited in water."

If microbial life had been present at the time, this environment would have sustained it. "Rock made of evaporite sediments from standing water offers excellent capability for preserving evidence of any biochemical or biological material that might have been in the water," said Squyres. "If you've an interest in searching for fossils on Mars," said Ed Weiler, NASA's associate administrator for space sciences, "this is the first place you want to go."

The fact that the hematite covered an area the size of Oklahoma indicated that a *large amount* of water had been involved. It was apparent from the light-toned rims that characterised the craters on the hematite plain that the outcrop in Eagle crater was regional. An inspection of a larger crater should give insight into the deeper structure of the plain, and fortunately there was such a crater 750 metres to the east, which had been named Endurance. It is interesting to consider that if the rover had landed far from any craters, it would have found the site littered with hematite spherules and yet, lacking any sense of the bedrock, would not have been able to provide any insight into their origin.

The lone rock
Although rocks were notable by their absence on the plain, there was one near the crater. Since this 35-centimetre-long 10-centimetre-high rock had been nudged by the airbags when the lander had rolled towards Eagle crater, it was named Bounce. "Not only did we score a 'hole in one', but on the way into the crater we struck the only rock around!" mused James Bell of Cornell, the leader of the pancam team. It was intriguing in its own right, for it was unlike any rock previously encountered on the planet, having, as Bell put it, "some shiny surfaces, almost mirror-like". A remote inspection by the mini-TES hinted at hematite, which, if true, would make this the first sample of hematite to be seen in *bulk* form. After driving to the rock on sol 62, the plan for sol 63 was pre-empted by a memory fault. Nevertheless, the rock was inspected by the microscope on sol 64, analysed by the spectrometers on sol 65 and overnight, drilled to a depth of 6.5 millimetres on sol 66, and the hole analysed overnight and the next day. By now, it was evident that the initial indication of hematite had been misleading – the rock had occupied so little of the instrument's field of view that the analysis had been biased by blueberries on the ground beyond.

As Bounce appeared to have cracked when the airbag rolled over it, on sol 68 Opportunity moved around to examine the opposite side of the rock, and then drove over it in an effort to break it apart; but to no effect. Intriguingly, the rock proved to be as unusual mineralogically as it was physically. Its chemistry implied that it was not excavated from the local bedrock, but must have been tossed in from another site. Nor was it similar to the volcanics Adirondack and Humphrey that Spirit had thus far examined on the plain at Gusev. Furthermore, it also differed significantly from the typical Martian basalt remotely sensed from orbit by the thermal emission spectrometer on Mars Global Surveyor. In fact, the 'best fit' mineral composition was 69 per cent pyroxene, 20 per cent plagioclase and 11 per cent olivine, which was

very similar to that of the Shergotty meteorites
– in particular EETA79001, which was found
in the Elephant Moraine in Antarctica in 1979
and was of a basaltic composition almost
indistinguishable from its terrestrial equivalent.
"There's a remarkable similarity in their
spectra," said Christian Schroeder of the
University of Mainz. Bounce could have come
from almost anywhere on the planet, but
infrared data on cooling rates after sunset
returned by the THEMIS instrument on Mars
Odyssey showed that a 25-kilometre-diameter
crater some 65 kilometres to the south had
splashed a 'ray' of rocky debris across the
landing site. "Some of us think Bounce rock
could have been ejected from this crater," said
Deanne Rogers of Arizona State University at
Tempe. If so, then it was a remarkable
discovery. An impact on such a scale would
have excavated to a depth of several kilometres, in which case Bounce may derive from
a volcanic plain underlying the Meridiani
deposit.

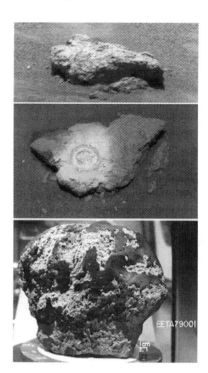

The rock named Bounce prior to and
after drilling, and (bottom) the
meteorite EETA97001.

EXPLORING ENDURANCE CRATER

On sol 70 Opportunity exploited the unobstructed character of the plain to dash 100
metres to Anatolia, one of several sinuous features that were barely resolvable in the
orbital imagery, and proved to be a chain of deep 'fractures' edged by rocks that
looked similar to those in Eagle crater. The rover spent two days navigating a safe
route across Anatolia, scraped a trench on sol 73, was stood down to enable its
software to be upgraded, and then, on sol 79, examined its trench.

The sinuous depression named Anatolia.

A view by Opportunity's forward hazard camera of the crater Fram, featuring the rock Pilbara, which was drilled and inspected by the microscope. On the horizon is the rim of crater Endurance.

On sol 81 the rover drove 32 metres, set a new record the next day by driving 141 metres, and on sol 84 reached Fram, an 8-metre-diameter crater some 450 metres from the landing site and 250 metres from Endurance. "Fram is really busted up," noted Squyres. In contrast to Eagle, this smaller crater had an ejecta blanket and a blocky interior. If Opportunity had landed in this crater, the task of interpreting the rock would have been considerably more difficult. On sol 85 the rover manoeuvred to a rock on the rim named Pilbara that had blueberries attached to the end of thin stalks. Evidently the berries were harder than the host rock, and as the matrix was eroded the berries remained in place on the protected stalks, until the stalks broke. The least-berried section of the rock was selected and, on sol 86, a 7.25-millimetre hole was drilled. When the microscope inspected the hole, there was an embedded berry exposed to view. On sol 88 the rover left Fram, drove 33 metres and halted to scrape a trench, which it studied over the next few days for a radial survey of the soil characteristics *en route* to Endurance.

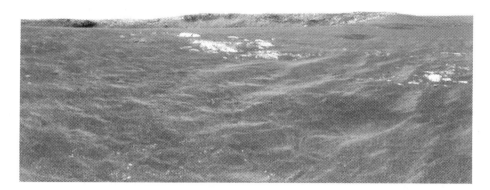

After leaving Fram, Opportunity drove towards Endurance and revealed intriguing structure in its far wall.

Being 140 metres in diameter, Endurance was much larger than Eagle or Fram, and the hope was that its walls would expose a sufficient thickness of rock to show the structure of the plain beneath the light-toned rock that had been studied to-date. By the end of sol 93 Opportunity was 70 metres from the near rim, and the view of the upper part of the far rim stirred considerable excitement, as it included a large mass of *dark* rock.

On reaching the crater two days later, Opportunity, tilted upwards at 5 degrees, crept to within 50 centimetres of the lip of the rim, beyond which the inner wall plunged at 18 degrees. After the navcam had made an initial 180-degree panorama of the interior, the pancam and mini-TES built up their own sequences at the best spatial and spectral resolutions. "This is the most *spectacular* view that we have seen on the Martian surface," Squyres pointed out, "not just for the science but for sheer beauty." The floor of the crater was 20 metres below the rim, and the 10-metre-thick

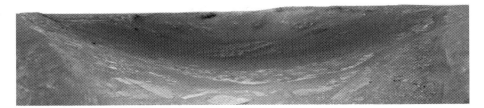

On reaching the western rim of Endurance, Opportunity took a panoramic view of its interior.

exposures in the wall displayed considerable layering. On the other hand, this was just a tiny fraction of the Meridiani deposit, which was several hundred metres thick. The mini-TES noted mineralogical variation in the wall, including basaltic, iron oxide and sulphate signatures. "The rock unit beneath that corresponding to the one we saw at Eagle crater looks fundamentally different from anything we've seen before," Squyres ventured.

The team now faced a dilemma, in that it was clear that to properly investigate the layering the rover would have to drill holes in it, yet if it did enter the crater it might not be able to escape it. The drivers considered the slopes of the northwestern and southwestern walls to be manageable but, with the onset of the southern winter, driving on the northwestern wall would tilt the solar panel on the deck of the rover away from the Sun. Hence, if the rover was going to enter the crater, it would have to be across the crater's southwestern rim. On sol 102, therefore, the rover began a traverse anticlockwise around the rim to assess this entry-point candidate, named Karatepe, where the bedrock on the scoured interior slope had been eroded to an almost even gradient averaging 25 degrees and looked like it had been 'tiled' by irregularly shaped plates. One complication was that the slope was littered with blueberries, which would act like ball-bearings and reduce traction. The streams of fines downwind of the berries indicated that they had been in position a long time, despite being on a slope.

On sol 104 Opportunity halted at a 30-centimetre-long 10-centimetre-high dark rock named the Lion Stone that looked, as Squyres put it, "like nothing we've ever seen before". Despite the uneven surface, the RAT was able to grind down the bumps and drill to a depth of 6.3 millimetres. As with the light-toned rocks in Eagle, it was finely layered, contained blueberries, and was rich in sulphur, implying that it had formed in water. "However," Squyres noted, "it was different in subtle ways – a little different in colour, a little different in mineralogy." The fact that the rock was on the rim suggested that it was from the deepest excavation. "It may give us the first hint of what the environment was like before conditions that created the Eagle crater rocks." The rover resumed its drive around the rim on sol 108, pausing from time to time to examine the soil and survey the ejecta. At a point on the southeastern rim it halted to take a second set of panoramas, then on sol 116 turned left and advanced to within 1.5 metres of the lip, beyond which the ground tilted away at 40 degrees.

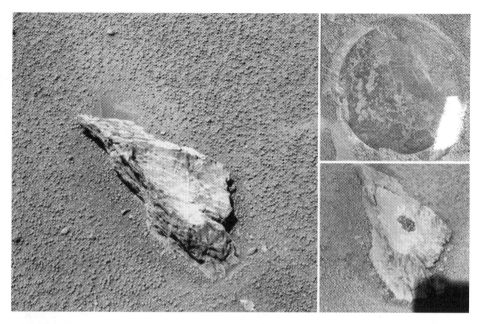

The Lion Stone on Endurance's rim was drilled and inspected by the microscope.

Having moved to the southeastern rim of Endurance, Opportunity took a second panoramic view of its interior.

"If we don't go close enough to the lip we can't see in," pointed out chief rover driver Brian Cooper, "but if we stray too close we could fall in!"

Opportunity next turned its attention to nearby rocks. "We're seeing rocks that have very interesting surface textures," said Wendy Calvin, a geologist from the University of Nevada at Reno. "These rocks appear to be from the same formation as the outcrop in Eagle crater – but with some differences." One, Diogenes, was riddled with the voids seen in the Eagle outcrop. The surface of another, named Pyrrho, displayed braided ripples that, on Earth, would arise if soft unconsolidated sediments were folded as they slumped into a shallow depression, which could only occur in a wet environment.

It was decided to have Opportunity enter the crater at Karatepe. After returning around the rim to this point, the rover manoeuvred up to the lip on sol 131, and the next day crept forward until its front wheels were just beyond the crest. On sol 133 it

The blueberry-strewn rock Pyrrho on Endurance's rim was found to possess a fine rippled texture.

advanced until all six wheels were over the lip, and then reversed out in order to assess the traction on the blueberry-strewn 18-degree incline. The following day it ran 4 metres downslope and then reversed 1.4 metres to further assess the traction. At least three layers were exposed on this part of the crater's wall, with distinct contacts. Designated alphabetically from the top, the succession of layers showed subtle variations in hue and texture. "Answering the question of what came *before* the evaporites is the most significant scientific issue at this time," Squyres said. "It could have been a deep-water environment; it could have been sand dunes; it could have been a volcano. Whatever we find out about that earlier epoch will help us to interpret the upper layer's evidence of a wet environment, and understand how that environment changed." The plan was to run down the slope and sample each layer. "We want to get to the contacts between the units, to see whether the environment changed gradually or abruptly," Squyres explained. Even if the lower layers were formed in dry conditions they might subsequently have been exposed to water, and so would offer clues to conditions prior to the drenching that formed the sulphate at the top of the stack.

A view from the southwestern rim of Endurance looking down across the Karatepe exposure that Opportunity was to try to sample.

As Opportunity advanced 3.2 metres on sol 135, the incline steepened to 20 degrees. After inspecting a tile in layer A about 15 by 36 centimetres whose shape led to it being named Tennessee, on sol 138 the RAT drilled a hole 8.2 millimetres deep. As expected, this was sulphate. However, the early ideas of the deeper layers were being reviewed: "We thought they were poorly consolidated sandy material," said Scott McLennan of the State University of New York at Stony Brook, "but as we get closer we're seeing more consolidated, hard rock." On finishing its analysis of Tennessee, the rover ran 70 centimetres further downslope to examine the first contact, and on sol 143 drilled a 3.8-millimetre hole named Cobblehill in layer B. The shallowness of this hole reflected concern that applying force to drill the rock might cause the vehicle to lose traction on the 23-degree slope, slide, and damage the articulation mechanism of the arm. On sol 145 it drilled a 4.3-millimetre hole named Virginia in layer C, and on sol 148 put a 4.5-millimetre hole named London in layer D. "This is the first detailed stratigraphic section ever done on another planet," pointed out McLennan. "We're doing exactly what a field geologist would do." In fact, this survey revealed unexpected similarities between the lower layers of rock and the layer corresponding to the outcrop at Eagle. While layers B, C and D were darker, they were sulphates ridden with blueberries, which established that a much greater depth of rock had been affected by the presence of water than had been believed. "I'd thought we might see just basalt below the salty top layer," said Squyres, "but it is salty as far as we've been able to see so far." Noting that the darker sulphates lacked voids and water ripples, he ventured that after having been laid down in water the material had been homogenised by "some kind of mixing", possibly by the wind stirring up the material during dry epochs. The clue lay in the ratio of chlorine to bromine. "At Eagle we saw a variation in the ratio of chlorine to bromine over distances of tens of centimetres. That's a clear indication of an evaporative process, since chlorides and bromides have different solubilities and they precipitate out under different conditions. But if you took that section at Eagle and mixed it all up, you'd see a uniform ratio of chlorine to bromine, just as we're seeing here."

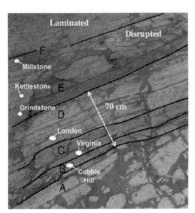

Opportunity found the Karatepe exposure to comprise a series of distinct layers (labelled alphabetically from the uppermost) which it proceeded to drill in turn.

As the layers were so closely spaced, the rover had been able to drill three of the holes from the same location. On sol 150 it reversed half a metre, turned left 40 degrees, and ran 1 metre down across the slope into a position from which to access a smooth section of layer E, where it drilled two holes: the first, named Grindstone, to a depth of 2.7 millimetres on sol 151 and the second, named Kettlestone, to a depth of 4.2 millimetres on sol 153. Although crossbedding was absent in layers B, C and D, it reappeared in layer E. "We're about 7 metres past the lip, as you'd draw a ruler across the ground, and we haven't found anything that isn't sulphate," reported Squyres. "So we have increased our estimate of the amount of sulphate by an order of magnitude compared to that at Eagle crater." As the rover reversed 1 metre on sol 155 to give the mini-TES a clear view of the recent holes, slippage on the berry-strewn slope increased to 11 per cent. When the vehicle advanced on sol 157 to gain access to layer F, the slope increased to 28.6 degrees and it ended up with one of its rear wheels off the ground. After manoeuvring to achieve a more stable position, it drilled a 6.3-millimetre hole named Millstone in layer F on sol 161.*

---

* See Colour Plate 7 for a false-colour perspective of the Karatepe drilling.

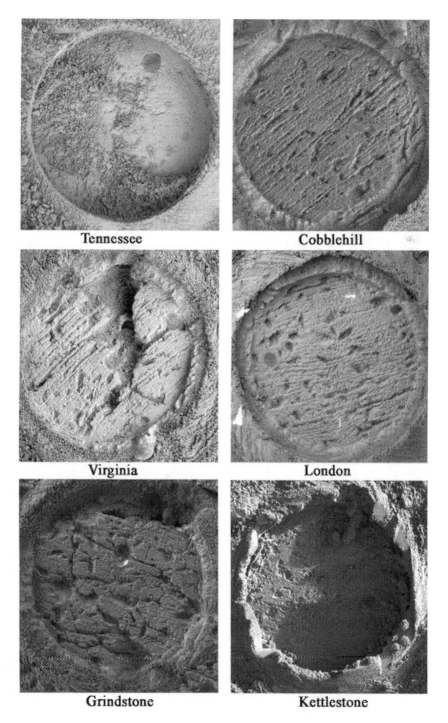

Microscope images of six of the holes drilled in the Karatepe exposure.

On sol 169 Opportunity resumed its descent into the crater over a patch with a 30-degree slope, at one point turning to manoeuvre cross-slope to the upper end of a number of narrow features with irregular edges that projected several centimetres from gaps between rocks. In driving down one of these Razorbacks, as the features were dubbed, its wheels snapped off pieces, and when it was noticed that one piece was within reach of the arm this was examined on sol 173. The fragment, named Arnold Ziffel, established that the Razorbacks formed when mineral-rich fluids ran through cracks in the rock, depositing minerals in veins which, by virtue of being hard, remained when the host was eroded away. This was yet further evidence of the history of water at Endurance. In reversing upslope on sol 175 to enable the mast-mounted cameras to view the site, the rover suffered 30 per cent slippage. Having ventured 13 metres from the lip and achieved such excellent results, it was decided to progress further into the crater.

The feature named Razorback.

On sol 177 Opportunity drilled a spot named Diamond Jenness, but the ground was littered with blueberries and the RAT had to clear a space to access the rock, the surface of which proved to be so irregular that in two hours it managed to penetrate only to a depth of 2 millimetres. When the microscope revealed this hole to contain a sliced blueberry, it was confirmed that the bedrock had been affected by water to a considerable depth. On sol 180 the rover reversed 1.5 metres, turned right, and drove down 50 centimetres to achieve a good position to continue the descent on a line that would avoid a patch of sand in which it might become bogged down. An accurate drive on sol 181 left it perfectly located to work on a target named MacKenzie, which it drilled to a depth of 8.4 millimetres on sol 182. Despite suffering 40 per cent slippage, the rover was able to advance another 8 metres on sol 184 to Inuvik, which was drilled to a depth of 7.7 millimetres two days later.

By now about 22 metres down the crater's wall, the plan was to turn right and run 20 metres across the slope to a rocky outcrop named Alex Heiberg, as the first leg of a traverse of about 80 metres along the lower slope of the southern wall to a point where Opportunity would be able to look *up* at the massive exposure of the Burns

From its vantage point low on the Karatepe exposure, Opportunity looked around the southern wall of Endurance towards the base of the Burns Cliff, and (bottom) the route that was planned for this traverse.

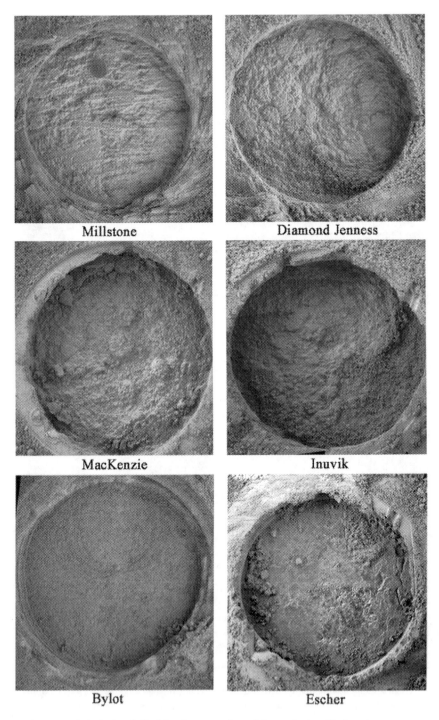

The second six holes drilled by Opportunity while exploring Endurance crater.

Cliff. To-date, the path had been on rocky tiles sprinkled with blueberries and sand, with strips of deeper sand between the rocks. In driving across the slope, the vehicle had tended to yaw from its intended route, and in trying to reach Alex Heiberg on sol 189 it lost traction and sideslipped downslope. Although the slope was just 17 degrees and the slippage was a modest 16 per cent, it achieved only 3.4 metres on sol 190. After accomplishing 5 metres on sol 191, ending somewhat upslope of Axel Heiberg, it turned on sol 192 and ran downslope. A rock named Bylot that was partially covered with darker sand was selected for inspection. There were some blueberries but they were less rounded, and when it was observed that some were cracked and there were normal-looking berries inside, it was speculated that after the hematite concretions had formed these had acquired an irregular outer coating. After a 7.6-millimetre-deep hole was drilled in Bylot, Opportunity reversed on sol 196 to give the mast-mounted cameras a look at a vein in the rock. On seeing that the vehicle's movement had broken off a 3-centimetre piece of rock from the vein, this was named Jiffypop and analysed, but when the RAT was directed towards it on sol 199 the tool's motor stalled. Diagnostics indicated that a piece of debris had become lodged in the grinder. While possible remedial action was being evaluated, it was decided to abandon Jiffypop.

On sol 201 the rover set off towards a tendril of sand that ran out from the field of dunes on the floor of the crater, but progress was slow, and when it began to look as if the wheels might bog down the decision was taken to return to Axel Heiberg, but the slippage while driving diagonally upslope deflected the vehicle 3 metres to the left of the target. It lost ground making a turn, and then on sol 206 struggled upslope to a rock in the Axel Heiberg group named Escher which, while not the one initially chosen, was accepted for examination. In the meantime, a visual inspection had established that if the RAT had been obstructed by a fragment of debris, the overnight thermal cycling had caused this to fall out.

After the solar conjunction in September, on sol 238 Opportunity manoeuvred to Ellesmere, the rock to which it had been heading prior to finding itself within arm's reach of Escher. The rocks in the Axel Heiberg group were distinguished by

The rock Escher, prior to being drilled.

their networks of cracks that resembled dried mud. "On seeing those polygonal crack patterns, right away we thought of a secondary water event, significantly later than the episode that created the rocks," said Grotzinger. Whereas Escher's surface was flush with the slope, there was a larger patterned rock ahead named Wopmay which was fully exposed. As the rover made a 20-metre drive diagonally down the lower slope on sol 249, slippage caused it to miss Wopmay by several metres. The

The rock Wopmay with the Burns Cliff in the background, and a close up of the rock's lumpy texture.

following day it began an indirect approach, and on sol 251 ended up so close to the rock that there was no room to deploy the arm! On sol 252 the rover backed off, and the pancam documented the rock utilising all its filters for maximum spectral resolution, repeating the procedure using the mini-TES the next day. The vehicle attempted to close in again on sol 255, but slipped and ended up 3.4 metres away. It managed to half this distance the next day, but this was not close enough for arm work. On sol 257, the rover set off on a vector that allowed for 50 per cent slippage, and finally attained a position from which it could reach the rock. Wopmay was about 1 metre in size, and the intersecting cracks gave it a lumpy texture. The slope and the loose material prevented the rover from gaining sufficient purchase to use the RAT, but the spectrometers were able to examine its surface. It was evident that the deepest rocks in the crater had been exposed to water *both before and after* the impact that created the crater, which suggested that water may have once pooled in the cavity.

On sol 262, having thoroughly inspected one side of Wopmay, the rover withdrew as a preliminary to seeking a position to use the RAT, but slipped off course, and the next day, suffering 100 per cent slippage, finished up sideways on to the rock. On sol 264 the manoeuvre was abandoned, and the eastward drive along the lower wall resumed with the objective of reaching a point from which the rover would be able to look up at the Burns Cliff. The plan was to 'tack' at 45 degrees, alternating upslope and downslope in an attempt to follow a given heading for this 21-metre drive, but by sol 271 Opportunity's drivers were having second thoughts. "We've made a careful analysis of the ground, and decided to turn around," said mission manager Jim Erickson. "To the right the slope is too steep – more than 30 degrees – and to the left there are sandy areas we can't be sure we could get across." This decision was not taken lightly, for 15 metres further on was an excellent vantage point from which to view where two of the layers in the cliff met at an angle. This feature was of great interest because the layers involved were below those that had been found to be sulphate. "This is a huge crossbed," explained Squyres, "the kind of thing that you'd expect from wind. We don't think these sediments were laid down flat and then tilted, we think they were laid down at an angle. When you see sediments laid down on a big scale, metres across, at an angle of 20 to 30 degrees, that says pretty unambiguously that it's a dune." However, the turnaround point was at the western end of the cliff, and over the ensuing several days the mast-mounted cameras took panoramas at the highest spatial and spectral resolutions peering obliquely up along the length of the exposure and, with the vehicle favourably oriented for solar power, the X-Band link to Earth was able to supplement the orbital relays in returning this data.* On confirming that the lowest layers of the cliff had been transported by wind, Squyres reported that this meant that Meridiani had been "not a deep-water environment but more of a salt flat, alternatively wet and dry".

Opportunity turned around on sol 295 and began to drive west around the wall of the crater. On sol 312 it swung left and started the long climb up the slope, and

---

* See Colour Plate 8 for a high-resolution view of the Burns Cliff.

The site where Opportunity's heatshield impacted and broke into several large fragments.

despite some slippage it crested the lip on sol 315 and advanced 2 metres down the outer slope, thus concluding a six-month exploration of the interior of Endurance. After recharging its batteries for several days, it set off towards its heatshield, 250 metres to the south. The impact had disturbed the surface, but had not penetrated deeper than a few centimetres. The structure had broken into two large pieces and many small fragments. As some of the debris might pose a hazard to the rover, the plan was to survey the site from a safe distance and select a route to inspect the larger pieces for an engineering assessment. A brief dust storm nearby increased the opacity of the atmosphere and diminished insolation by 30 per cent over the period sol 329 to sol 331. This meant that the rover had to minimise its power consumption, but it later closed in on one of the larger pieces and on sol 341 the microscope made a 96-frame mosaic. "We're examining the images to determine the depth of charring in the material," noted Christine Szalai, a spacecraft engineer at JPL. Several days later, Opportunity was sent 10 metres north to inspect a basketball-sized rock with a deeply pitted surface that remote-sensing by the mini-TES had shown to be rich in metals. In fact, this Heatshield Rock, as it was named, proved to be mostly iron and nickel, and was a meteorite. "This is a huge surprise," said Squyres, "although perhaps it shouldn't have been." Such a meteorite represents the interior of a large planetesimal shattered by a giant impact. Noting that Mars should have been hit by many more rocky meteorites than iron ones, Squyres suggested that some of the cobbles on the plain might be rocky meteorites. Such interlopers would be hard to spot on any other part of the planet, but on the bland plain at Meridiani they drew attention to themselves, in the same manner as did meteorites on the ice fields of Antarctica.

Heatshield Rock.

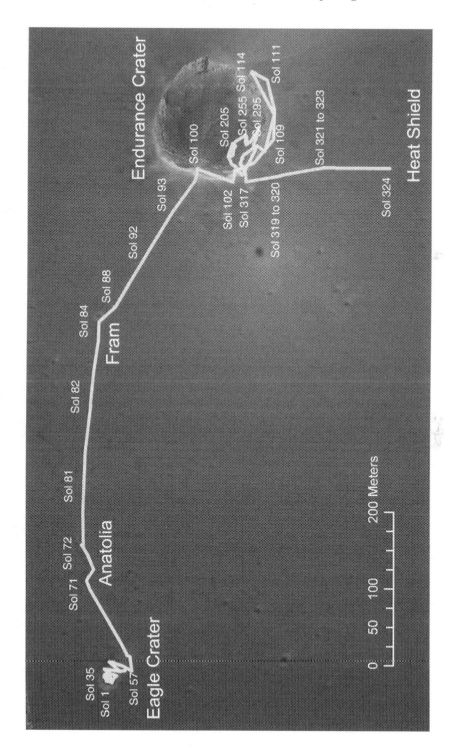

A summary of Opportunity's traverse to sol 324.

DRIVING SOUTH

Having provided considerable insight into the bedrock of Meridiani Planum by its exploration of Endurance, on sol 358 Opportunity set off south for the nearest 'etched' terrain, several kilometres away, which seemed to mark where other strata of the bedrock were in outcrop. Although now well past its nominal mission, both in terms of duration and distance, the vehicle had survived the winter and was still in excellent condition. "The terrain that we're crossing is so flat," explained route planner Frank Hartman, "we can see a long way ahead." Ironically, on attempting to cross a dune of windblown sand and dust on sol 446, the vehicle became bogged down. On sol 458, after 190 metre's worth of almost tractionless wheel rotations, it managed to cover the few metres needed to withdraw from the trap that had been dubbed the Purgatory Dune. After creeping back to the edge of the dune in order to examine the trenches left by its wheels, Opportunity withdrew in search of a safer route to its next objective, the crater Erebus, some 500 metres further south on the etched terrain which, with luck, would shed additional light on the stratification of the plain.

REFLECTIONS ON MERIDIANI

In a paper in *Nature* in September 2004, Brian Hynek, who had analysed the data from Mars Global Surveyor and Mars Odyssey that implied Meridiani Planum was a pyroclastic blanket, reported a follow-up investigation on the basis of the ground truth from Opportunity. He traced the bedrock outcrops many kilometres north, east and west of the landing site and concluded that there had been "a body of water at least 330 square kilometres" – comparable in area to the Baltic Sea. In THEMIS data, a high thermal inertia in the hours after sunset meant rocks, and a low thermal inertia meant fine-grained material. "The thermal inertia for this area is relatively high, indicating that it contains substantial bedrock." Hynek exploited altimetry to chart the topography, and high-resolution imagery from Mars Global Surveyor to inspect the nature of the surface. "By examining the landing site area from orbit, I was able to see what signatures show up in the thermal data, as well as in the visible imagery." In essence, he had traced the areal extent of the bedrock that had been seen at the landing site. His first discovery was that the 'etched' terrain, whose morphology had caused it to be distinguished from the smooth dark plain in the mapping process, was actually the bedrock beneath the hematite. In retrospect, it was apparent that while hematite had drawn attention to the site, it was the sulphate that required to be traced.

"There is still an unresolved issue," warned Michael Carr of the US Geological Survey at Menlo Park in California, "and that's where is the basin in which all this occurred?" Although Hynek did not identify a topographic boundary, he noted that his paper was a progress report and that further investigation might greatly expand the area sufficiently to reveal the enclosing landform. Tim Parker of the University of Southern California, who first suggested that there might have been an ocean in

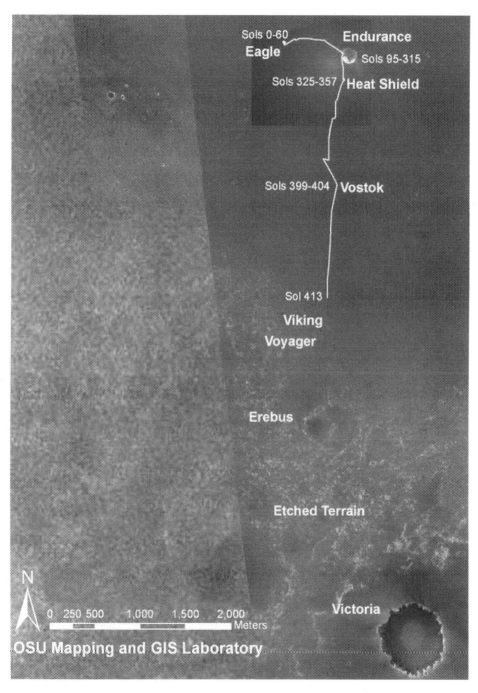

A summary of Opportunity's traverse to sol 413, showing its route towards the 'etched' terrain to the south.

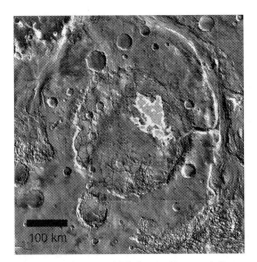

The crater Aram, which has a disrupted floor and an exposure of grey hematite (shown here in false-colour overlay).

the northern lowlands, had long argued that until the full extent of the sedimentary rock was determined it was premature to dismiss the possibility that Meridiani had once hosted a considerable amount of surface water. Finding that the aqueous fluid had been *acidic* offered a resolution of the 'carbonate paradox' that represented the main issue for the hypothesis that Mars had once been warm and wet. In a paper in *Nature* in September 2004, Alberto Fairen of the University of Madrid, who had studied the Rio Tinto in Spain – which has a chemistry and mineralogy similar to Meridiani – ventured that Mars had an early ocean that "probably lasted nearly 1 billion years". Acidified by iron and sulphur, this ocean would have had a very different chemical evolution to that of terrestrial oceans. In particular, it "left no Earth-like sediments, such as carbonate minerals" as carbonates would not have been able to precipitate out of an acidic solution. Although Meridiani lies outside the shorelines suggested by Parker on the basis of his study of the Viking imagery, he has recently pointed out that (1) the altimetry data from Mars Global Surveyor charts a scarp that traces a ragged line featuring embayments and fjord-like indentations that he interprets as a shoreline at approximately the datum, and (2) despite being a 'high plain', Meridiani Planum lies 2 kilometres *below* this elevation. Interestingly, a smaller patch of grey hematite exists in Aram, a nearby crater with a diameter of 280 kilometres whose floor displays the kind of collapse associated with the chaotic terrains, and because the morphology suggests that this hematite was buried and is being exhumed, it might be much more extensive than is indicated by its surface expression. If this is so, then there may indeed have been a northern ocean that from time to time inundated the low-lying landscape of Arabia Terra.

9

Future prospects

WATER AND LIFE

The 1960s premise in setting out to search for life on Mars was to ask, in essence, what kind of terrestrial life was best suited to the conditions on the surface of that planet (which were not well known) and then to presume that life there would be similar. At that time, the 'dry valleys' of Antarctica were the best Earthly analogue for Mars. Since then, many niches that had been considered sterile have been found to host ecosystems. In fact, extremophiles are metabolically diverse, and capable of utilising almost *any* chemical energy that is abundant. When Viking was designed, it was believed that all life ultimately derived its energy from sunlight, that metabolism involved gaseous exchange of carbon dioxide, and that wherever the biology package was placed on Earth it would detect life. However, it is clear in hindsight that it would *not* have detected many of the extremophiles. A genetic study has shown that the 'common ancestor' for terrestrial life was a hyperthermophyllic anaerobic autotroph that used hydrogen. The lesson from Earth would therefore seem to be that life is the direct result of chemical evolution, and developed as soon as it became possible for it to exist. As Bruce Jakosky, an astrobiologist at the University of Colorado, has put it, "the environmental prerequisites for life include only the presence of liquid water, access to the biogenic elements, and a source of energy that can drive chemical disequilibrium; these are not terribly stringent requirements". When terrestrial life originated, there was intense volcanism, liquid water was on the surface and the air was predominantly carbon dioxide. If, as seems likely, early Mars was similar, then life may well have developed there, too. In accordance with the 'follow the water' strategy, the aim of the Mars Exploration Rover missions was to seek incontrovertible evidence that there was once a significant amount of liquid water on the surface, and this was proved. Furthermore, the 'carbonate dilemma' appears to have been resolved, because the volcanism would have made the water highly acidic. As Ed Weiler, NASA's associate administrator for space sciences, had previously explained to reporters, the missions were "not about rocks, but about the potential for life on another planet", and, as demonstrated by the Rio Tinto in Spain, a microbial ecosystem can exist in a sulphuric acid environment. If life still exists on Mars, it is likely to be an anaerobic autotroph living in a hydrothermal

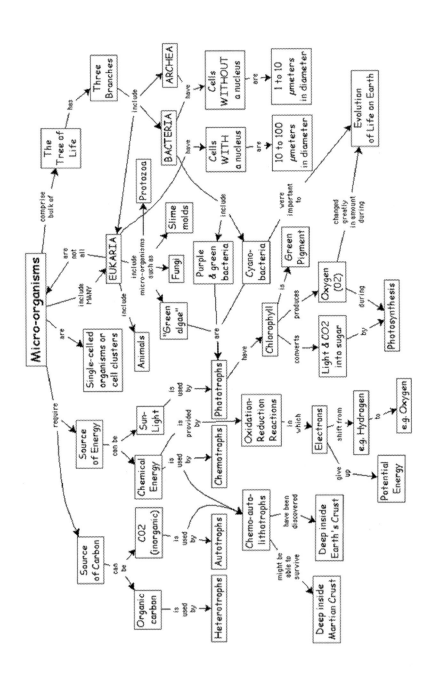

An outline of the rationale for seeking life on Mars.

system. Vic Baker of the University of Arizona, who has argued that "early Mars provided an arguably better habitat for the inception and incubation of early life than did Earth", thinks there may be "a deep subsurface containing methanogenic archea". The next lander sent to test for life will probably be equipped with a drill to enable it to obtain samples from a considerable depth! However, in view of the disputed status of the oldest candidates for terrestrial stromatolites, if we find structures on Mars that *could be* fossils, it may be impossible to *prove* their biological origin.

SNIFFING FOR LIFE

One way of detecting the presence of life on Mars would be by gaseous emissions from metabolism. Mariner 9 set an upper limit for methane in the atmosphere of 20 parts per billion by volume; the figure for Earth is 1,700 ppb. The first reliable detection was reported in September 2003 at the Division of Planetary Sciences of the American Astronomical Society by Michael Mumma, director of the Center for Astrobiology at the Goddard Space Flight Center in Maryland. In March and May 2003, Mumma had undertaken spectroscopic observations using the 8-metre Gemini South Telescope on Cerro Pachon in Chile and NASA's 3-metre Infrared Telescope Facility on Mauna Kea in Hawaii, and detected 10 ppb, averaged across the globe. Shortly thereafter, Vladimir Krasnopolsky of the Catholic University of America in Washington DC, reported a similar figure based on data secured in 1999 using the 3.6-metre Canada–France–Hawaii Telescope on Mauna Kea. Such observations were difficult, because the signature from the methane in the Earth's atmosphere had to be subtracted. When Mars Express entered orbit around Mars in January 2004, it had a spectrometer capable of detecting methane and other trace gases. Vittorio Formisano of the Institute of Physics and Interplanetary Science in Rome, Italy, leading the team, said in March that averaging the data from several orbits in January and February gave a global concentration of methane of 10 ppb, rising to 40 ppb in some areas. However, for the first few months the observations were complicated by dust in the upper atmosphere resulting from the regional dust storm in December 2003. At the International Mars Conference at Ischia in Italy in September, Formisano said that the high methane concentrations overlapped the areas in which water vapour was concentrated – above 10 kilometres water vapour was well mixed, but nearer the surface its concentration was several times greater in three equatorial areas: Arabia Terra, Elysium Planitia and Arcadia–Memnonia. These matched the areas in which the neutron spectrometer on Mars Odyssey had inferred the presence of hydrogen, suggesting a common underground source. The hydrogen had been interpreted as the presence in the near-surface of either hydrated minerals in rocks or water-ice in the soil. If it was ice, might this be geothermally heated water that migrated towards the surface and then froze and, if so, might it sustain a subsurface microbial ecosystem that was releasing gases into the atmosphere? Once the atmosphere had cleared, Mumma found a strong signature of methane, and told the November meeting of the Division of Planetary Sciences of concentrations of 250 ppb at Valles Marineris and Syrtis Major, compared to 20–60 ppb at higher latitudes. "I'm shocked by this result," he admitted. "At these two points on Mars, the data imply that there were significant methane releases." Krasnopolsky and Formisano had observed a single spectroscopic feature, but Mumma had measured two lines and achieved a signal-to-noise ratio of 20. "This cannot be waved away as measurement error," he emphasised.

If the methane had been released in a single brief emission, or as the result of a cometary impact, then the winds should have distributed the gas globally, giving a uniform concentration. The lifetime of a methane molecule in Earth's atmosphere is about 10 years before it is dissociated by solar ultraviolet to water and carbon dioxide. Although the thin Martian atmosphere is less able to block ultraviolet, the lifetime of methane is about 350 years because that planet orbits further from the Sun and the insolation averages 40 per cent of the Earth's. The uneven distribution indicated that the gas was being continuously released from a number of distinct sources. Although methane can be generated by volcanic processes, most of the methane on Earth can be traced back to microbial activity. Terrestrial methanogens are exclusively archea. The amount of methane detected in the Martian atmosphere could readily be produced by volcanic outgassing at a rate of 150 tonnes per year. While there are lava flows that appear to be only several tens of millions of years old, remote-sensing from orbit has yet to identify any hot spots to indicate ongoing activity. By setting an upper limit of 0.5 ppb for sulphur dioxide, Krasnopolsky established that the most abundant volcanic gas on Earth is spectacularly absent on Mars. But if geothermal heat was melting methane-ice (clathrate) and releasing the methane molecules that were locked in the crystals of water-ice, the source would not necessarily correlate with volcanism. If this was so, the methane, irrespective of how it was made, would be ancient. Another possibility was that water in aquifers beneath the permafrost was oxidising minerals in basaltic rock, a process known as 'serpentisation' because it turns basalt into serpentinite, and the liberated hydrogen was combining with carbon dioxide to create methane, which seeped out of the surface into the atmosphere.

At the conference at the European Space Research and Technology Centre in Noordwijk in the Netherlands in February 2005 to review the significance of the first year's results from Mars Express, Formisano said that the Planetary Fourier Spectrometer had detected formaldehyde at a concentration of 130 parts per billion. He argued that this was due to iron oxides at the surface turning CH_4 to CH_2O. Because formaldehyde would be dissociated within a matter of hours, this meant that more methane was being produced than was directly detectable – several million tonnes per year, which was much more than could be expected from outgassing. Most of the methane was being turned into formaldehyde near ground level, not dissociated by ultraviolet. The more the amount of methane inferred to be present exceeded the amount that could reasonably be explained in terms of non-biological processes, the stronger the case for a biological origin. "I believe that until it is demonstrated that non-biological processes can produce this, possibly the only way to produce so much methane is life," pointed out Formisano. "I do *believe* there is life inside the planet, maybe 50 to 100 metres below the surface, but there is a long way to go to demonstrate that." Unfortunately, Beagle 2, which had carried a spectrometer that might have resolved this issue, had failed.

WHAT NEXT?

Over the decades, more robotic probes have been launched towards Mars than any other planet. The more that is learned of the planet, the greater our fascination with it. As each mission answers some questions, it raises others that new missions, with new instruments, will seek to answer.

Mars Reconnaissance Orbiter

Scheduled for launch in August 2005, NASA's Mars Reconnaissance Orbiter is to enter a near-polar elliptical orbit in March 2006, and spend six months aerobraking into its operating orbit. As the largest orbiter yet to be sent to Mars, its instruments will enable findings from earlier orbiters to be followed up. Specifically, it will:

- use the most powerful camera ever placed into orbit around the planet to image selected sites at 1-metre resolution, in particular where there is stratigraphy in outcrop or where the terrain has been shaped by water erosion;
- use a spectrometer with a higher spatial resolution than previous instruments to investigate selected sites in search of minerals diagnostic of water;
- use an improved form of the infrared radiometer flown (and lost) on both Mars Observer and Mars Climate Orbiter to profile the atmosphere;
- use an improved form of the colour-imaging system on Mars Climate Orbiter to monitor the weather on a daily and seasonal basis, in order to further define the present climate;
- use a sounding radar to probe to a depth of several hundred metres in search of subsurface structures, particularly in the polar regions;
- have a 3-metre-diameter high-gain antenna for a data-rate 10 times that of any previous orbiter;
- examine potential landing sites for future missions; and
- serve as an orbital relay.

Phoenix Mars Scout

After the loss of Mars Polar Lander in 1999, a sister craft scheduled for 2001 was initially cancelled and then reassigned as the first mission in the Mars Scout series. Continuing the 'follow the water' strategy, the Phoenix lander will be launched in August 2007 and land in May 2008 in the northern hemisphere just outside of the seasonal cap, where the neutron spectrometer on Mars Odyssey showed there to be near-surface ice. It will have improved versions of some of the instruments lost on Mars Polar Lander, and others from the 2001 mission. A 2-metre-long arm will dig a trench to supply subsurface samples to an instrument to measure the 'volatiles', in particular water and carbon dioxide physically and chemically bound in the soil, and meteorology instruments will then undertake long-term monitoring of the local

An artist's depiction of Mars Reconnaissance Orbiter.

An artist's depiction of the Phoenix lander.

environment. In a paper in *Icarus* in 1990, Lynn Rothschild suggested that microbes might live near the polar caps, where the peroxy compounds will have been broken down by the presence of water. If there is extant life on the planet, it might well be just beneath the surface in such a location.

Mars Sample Return

The next major milestone will be to retrieve a sample from the Martian surface and return it to Earth for laboratory analysis. However, sample-return appears to one of those frustrating missions that is always 10 years into the future! Furthermore, the greater the likelihood of there being extant life, the greater will be the outcry about the potential risk of a returned sample contaminating the terrestrial biosphere, and it is entirely possible that a rigorous application of what has come to be known as the 'precautionary principle' will preclude such a mission.

Index